100 Ways
To Future-Proof Your Brain:
In the Age of AI

Sarah Baldeo

ISBN: 978-1-0695015-0-9
Copyright © 2025 Sarah Baldeo
References updated and expanded, 2026.
All rights reserved.
No part of this publication may be reproduced without permission.
Published by Arc & Edge
Toronto, Ontario, Canada

First edition

Dedication

With Love and Thanks.

No dream this big is ever truly solo.

To my husband, Iskandar - every time I chase some magnificently gigantic goal, you're right there beside me, unwavering. Thank you for making space for every version of my ambition, and for loving the whole storm of it. Thank you for reminding me to rest and showing me that I am worth so much more than my achievements – that I am loved even when I think I am failing at everything.

To my son, Ethan - thank you for your patience through years of "Mom's working." I know there were video games we could've been playing together. Thank you for being an amazing son over a decade of it being just you and I against the world. Your understanding gave me the room to build something that, I hope, will someday inspire you, too. Also, congrats on your appointment as your Moms Book Tour Assistant.

A special thank you to my grandmother Lois - who was the safe harbor for me always, who taught me strength, resilience, and constant kindness. My late grandfather Herbert who was a lifelong entrepreneur and reminded me to laugh in every lesson he imparted. My grandmother Doreen, as your only granddaughter you've always encouraged me to be the absolute best, to exceed every expectation – even yours. To my late grandfather Dr. Isaac Baldeo, who inspired this love for neuroscience and psychology in his only granddaughter at a young age – I miss you everyday and our philosophical debates. To my Aunt Ruth, who I was named for as Sarah Ruth – thank you for cheering on my big dreams, from near or far.

To my bonus mom, Vera, stepmother isn't the word for 30+ years of raising me - you taught me the art of resilience, of surviving the hard stuff with lipstick on and head held high. Your lessons in strength started young and still ring loud. To Lou - you were my stepfather for over a decade, and you encouraged my big dreams and writing always – that bond is never broken. To Carmita, my former mother-in-law - your belief in me never wavered, not even after the chapters of life shifted. That kind of grace is rare, and I hold it close. You will always be my family and the best grandmother I could wish for Ethan.

To my mother-in-law, Afat - your love and acceptance have wrapped around me like a second skin. Thank you, Dad and Ismat for making me feel like family with no hesitation.

To Celeste, my sister and my chosen family - you've loved me through every iteration of who I am and who I might still become. We have known each other for a lifetime, and it's no coincidence being born 8 days apart. How many journeys have you committed to over a 100-stop book tour!

This book is filled with science, stories, and strategy - but between every line is a village - a much larger one than any dedication pages can hold. To each person, each champion of my work - thank you for being the village of my life.

- Sarah

About the Author

An experienced AI neuroscientist, global futurist, corporate strategist, and 6x Founder with 2 exits, Sarah Baldeo is a recurring TED Speaker. In 2026, she reached 20 years of professional keynote speaking across Canada, the US, Europe, and Asia, surpassing her 1100th stage in December 2024 with the Department of National Defense. Sarah is a former member of the prestigious Canadian Association of Professional Speakers and an active member of the National Speakers Association. Nominated for the DMZ 2024 Women of the Year Award, Top 25 RBC Women of Influence Award, the 2024 Entrepreneur of the Year Award, and a finalist for the 2025 Women in AI Social Impact Award, Baldeo's research and work centers on the neuroscience of change and resilience.

Baldeo has been a recurring neuroscience expert on the brain and behaviour on more than a dozen live news and national networks such as CTV News in Canada and NBC News in the USA. 2025 began with her hit segment "The Neuroscience of Habit Formation" on nationally acclaimed daytime TV Show CTV The Social and her Keynote in Kuala Lumpur at the 5th Annual Digital HR Transformation Conference: "The Neuroscience Equation for the Future of Work." In 2025, Sarah was featured in the Globe & Mail as part of the "Reimagining Wealth" series, where she speaks about building a future-proofed brain and life.

As a female executive and first-generation Canadian of Jamaican and Guyanese descent, her story wasn't all wins; Baldeo's story of survival through divorce, lone parenting, cancer, multiple surgeries, gun violence and severe physical attacks connected with audiences in her 2023 TED Talk: The Neuroscience of Resilience, and her 2024 TED Conference guest appearance speaking on the "Chemistry of Emotions." Her keynotes explore the intersection of neuroscience and technology. The central ethos of her research are physiological changes to the human brain and how cognition of these types of adjustments can be a mechanism to move beyond survival to living a full and ambitious life, as well as enabling change management at some of the largest Fortune 100 companies on the planet including Coca-Cola, Sony Ericsson, and the Ontario and Florida Governments.

Currently CEO at ID Quotient, a neurotechnology firm founded in 2010, Baldeo has held senior executive roles at Canada News Wire, Ceridian, FinancialForce, Gartner, Insurance Search Bureau, Deloitte Canada, and CGI. Her academic background includes an Executive MBA via University of Toronto, a neuroscience degree from York University, and certifications from Princeton, Toronto Metropolitan University, and University of Illinois. Her PhD Research resides at Middlesex University in the UK where her team focuses on MRI and behavioural research studies on how the brain's physiology evolves through the use of technology and AI tools. In March 2024, Baldeo's neurotechnology firm was awarded the WEOC Federal Expansion partnership to grow into the USA and Europe, expanding their mission of democratizing access to STEAM resources.

Sarah's board roles have included being a mentor at Women's Infrastructure Network & Deloitte Canada, Board Advisor at UCLA Women in Leadership, Board Member at Cybersecurity Global Alliance, Executive Advisory at Civic Action Emerging Leader's Network, Chairwoman of the Canadian Professional Background Screening Association, and the Women in AI USA Illinois State Ambassador (2024-2025). Presently, she lives multi-city in Chicago and Toronto with her family, including her husband and teenage son. Her work remains focused on integrity, intellect, and democratizing scientific and technological knowledge to all – something she credits her own self-made success to; the ability to pursue lifelong education and embrace constant curiosity.

Author Note To The Reader

The human brain is not merely the epicenter of cognition; it is the crucible of all lived experience, the original technology from which every other tool - physical, cultural, digital - has emerged. It is simultaneously archaic and astonishing, stubborn in its evolutionary pacing yet capable of the most exquisite neuroplastic feats. If there is to be a next stage in human development, biologically or sociologically, it must begin here.

Historically, every seismic shift in how we live - from the Agricultural Revolution to the rise of industrial capitalism to the current algorithmic epoch - has been paralleled by shifts, however incremental, in how we think. Our neural architecture, forged for survival on Paleolithic plains, has been forced to adapt to increasingly abstract environments: first in domesticated settlements, then in mechanized economies, and now in hyper-networked digital realities. And while our surroundings have evolved at an exponential rate, our biological systems, including the brain, have lagged - optimized for scarcity and threat, not abundance and information saturation. We hold fast to our lizard brain in times of stress.

To understand the implications of this lag, one must dissect the broader history of human behavioural inflection. The end of the Pleistocene Epoch catalyzed the dispersal of Homo sapiens across varied ecologies, each shaping distinct social and cognitive patterns. Climatic anomalies - such as the Medieval Warm Period, which altered agricultural viability across continents - recalibrated both population distributions and civilizational priorities. The Holocene Climatic Optimum ushered in unprecedented environmental stability, enabling the emergence of complex societies - and with them, novel neural demands related to language, ritual, and planning.

Yet in the 21st century, it is not tectonic plates or temperature gradients that most rapidly alter human behaviour - it is technology. And the question before us is urgent: can the neurobiological substrate of Homo sapiens evolve, or at least recalibrate, to keep pace with the accelerating demands of an age defined by AI, ambient surveillance, synthetic stimuli, and cognitive overload?

Despite dramatic advances in neuroimaging and cognitive science, we still comprehend only a fraction of the brain's full operational schema. The oft-cited myth that we "only use 10% of our brain" may be reductive, but the reality is more sobering: we have mapped only an estimated 10% of the brain's functional systems with clarity - and we understand even less about how to architect deliberate change within that systems architecture.

And while neuroscience is unlocking the secrets of synaptic transmission, glial function, and neurochemical regulation, we are simultaneously witnessing a worldwide crisis of cognitive resilience. The World Health Organization estimates that nearly one billion individuals globally live with a diagnosable mental disorder - a statistic that transcends geography, ideology, and income bracket. In regions ranging from Scandinavia to Sub-Saharan Africa, the rising tide of chronic stress, trauma exposure, and attention fragmentation is rewriting both public health policy and private existential experience. UNICEF reports that over 75% of children globally endure at least one adverse experience - often before the neural circuits responsible for emotional regulation have fully developed.

These phenomena are not tangential to the project of progress - they are central to it. As we cross the threshold into what many now call the Sixth Industrial Revolution - defined not by steam or silicon, but by the convergence of artificial intelligence, neurotechnology, and bio cognitive enhancement - we are confronted with a rare imperative: to intentionally evolve the very organ that interprets, creates, and responds to reality. Therein lies the provocation - and the promise - of *100 Ways to Future-Proof Your Brain*.

This book is not a fluffy tour of the brain's curiosities. It is a deliberate synthesis of empirical neuroscience, behavioural design, and applied neuroplasticity, written to equip readers with the frameworks and techniques required to reclaim authority over their cognitive evolution. Based on my functional magnetic resonance imaging (fMRI) research, informed by a constellation of peer-reviewed findings from global neuroscientific institutions, and intersected with 2 decades of creating and building technology for how the brain navigates interfaces, this work offers not metaphorical guidance, but mechanistic tools.

Readers will navigate a labyrinth of ideas, some of which challenge everything we think we know about neural evolution:

- The strategic redistribution of neural activity away from the limbic system - the seat of inherited survival responses - and toward the prefrontal cortex, the hub of creativity, foresight, and rational modulation.
- How maladaptive stress responses, originally protective, have become pathologized in modern life - and how to un-map these pathways using targeted cognitive interruption.

- A methodology I refer to as **ballistic interruption**: a form of patterned, volitional disruption that facilitates new neural pathways through choreographed unpredictability.
- Why novelty, ambiguity, and exposure to cognitive dissonance are not threats, but vaccines - inoculations that protect the brain from rigidity and decline.
- How cultivating intentional neural friction - that is, choosing challenge over comfort - accelerates adaptive plasticity and preserves executive function in an age of automation.

Contents

Introduction 1

Way 1: Fight Your Instincts to Keep Going! 4

Way 2: Challenge Your Brain's Prime Directive – Go Beyond "Stay Alive!" 7

Way 3: Understand The Chemistry Behind Your Reactions! 9

Way 4: Understanding That Trauma is Generational AND Genetic . 11

Way 5: Rewiring Your Instincts – from Limbic System to Frontal Cortex 13

Way 6: Cognitive Flexibility & Being Your Own "Interior Designer" 15

Way 7: This is the Sound of Silence 17

Way 8: Interrupting Those Automatic Negative Thoughts 19

Way 9: The Novelty Pill 21

Way 10: Active Problem Solving – no AI-assistance! 23

Way 11: Interrupting Panic with Logic – Rewiring your Survival Response 25

Way 12: Be Willing to Re-align Your Memories 27

Way 13: Challenging The Fairytale Archetypes 29

Way 14: Don't Stop Evolving 31

Way 15: CHOOSING Change vs. Letting it Happen 33

Way 16: Embrace the Weird, the Bizarre, the Discomfortable! 35

Way 17: Consciously Switch Environments 36

Way 18: Get Off Screens! 38

Way 19: Break The Toxic Relationship with Endurance 40

Way 20: Neural-Trigger Mapping 44

Way 21: Empathy – The Super Power 46

Way 22: The Copycat 49

Way 23: Navigating Health Crises 51
Way 24: Re-architecting Self-Perception Through Job Loss 53
Way 25: Make Peace Your Preferred State of Being 55
Way 26: Hack Your Human Blueprint 57
Way 27: Build Your Bionic Brain 59
Way 28: Match Your Brain to Your Environment 61
Way 29: Do it Yourself 63
Way 30: The AI Study Buddy/Therapist? 66
Way 31: Journalling, By Hand 68
Way 32: Be Multi-Source 70
Way 33: Go Cold Turkey with Tech Sometimes 72
Way 34: Handwriting & Cursive – Bring it BACK! 74
Way 35: Say Thank you – Out Loud! 76
Way 36: The 4x4 Method…No it's not Algebra 79
Way 37: Resilience Playlists 81
Way 38: The Gratitude Anchor 83
Way 39: Habit Formation – but Make it Seasonal 85
Way 40: Pumpkin Spice and Everything Nice 87
Way 41: The Snow is Coming 89
Way 42: Coachella Summers and Beyond 91
Way 43: Take a Hike…seriously. 93
Way 44: Hotdesking for Your Hippocampus 95
Way 45: Draw-Storming 97
Way 46: Around the Campfire ? 99
Way 47: Space – the Final Frontier? 101
Way 48: The Castle 104

Way 49: Alpha & Omega 106

Way 50: Walking….Really it's that simple 108

Way 51: HiiT 111

Way 52: It's all about Balance 113

Way 53: Throw a Dance Party! 115

Way 54: Explore outside Your Tribe 117

Way 55: Go Clubbing 119

Way 56: The Imitation Situation 121

Way 57: Mentor 123

Way 58: The Mirror Project 125

Way 59: Tech Webinars 127

Way 60: Learn to CODE 129

Way 61: Human First, Tech Second 131

Way 62: Anti-Google 134

Way 63: Device-Free 136

Way 64: A Staycation isn't the same as a VACATION 138

Way 65: Paraskevidekatriaphobia – that's a mouthful! 140

Way 66: 8 hours a day keeps the… 142

Way 67: SLEEP…Quality over Quantity 144

Way 68: As the song goes… "Must be Love on The Brain" – Rihanna 146

Way 69: Make Uncertainty Your Training Ground 148

Way 70: Sunny with a Chance of Rain, Clouds, and Thunder? 150

Way 71: Learn how to do the Things You Think You're Bad At! 152

Way 72: Build a Relationship with Friction 154

Way 73: The True Cost of Perfectionism 156

Way 74: Protect Your "Novelty" Drive 157
Way 75: Buy Yourself Flowers…Like Miley Cyrus Says 159
Way 76: Step Inside a Smarter Space 161
Way 77: Choose Luxury Over Scarcity 163
Way 78: Understanding The Aging Brain – By Gender 165
Way 79: Ditch The Attention Economy 167
Way 80: Get Lost at a Street Market 170
Way 81: Visit a Sacred Space that Perhaps Isn't Aligned to Your Beliefs 172
Way 82: Swap the Org Chart – UPSIDE DOWN 173
Way 83: Be Alone, ON PURPOSE 174
Way 84: Light a Campfire 176
Way 85: Eat the Carbs, Feed the Brain 178
Way 86: Have a Glass of Champagne 180
Way 87: Get Outta TOWN! 182
Way 88: Make a List. It's That Simple. 184
Way 89: Stop Doom-Scrolling The Apocalypse 186
Way 90: Try Eating Once a Day 188
Way 91: Make Coffee a Ritual, NOT a Reflex 190
Way 92: Get Good at Rejection! 192
Way 93: Train your Brain to Notice and UNDERSTAND Beauty 194
Way 94: Feel the Temperature Shift 196
Way 95: Get Closer to Water 198
Way 96: Learn to be Still in a World That Never Stops 201
Way 97: Discover Your Inner Picasso 202
Way 98: Build a Brain That Belongs To YOU 204

Way 99: Reclaim Wonder, Deliberately 206
Way 100: Choose Neuroplastic Hope 208
Conclusion - Future Proofing the Brain is Future Proofing YOU 210
References 212
Notes 218

Introduction

The brain you're using right now wasn't built for a world of generative and agentic AI, robotic automation, quantum computing and digital acceleration. It's running ancient software - wired for survival, not for the speed, complexity, and constant disruption that define our lives today. For the first time in human history, technology is outpacing the human capacity to adapt in real time. Not because we're lazy. Not because we're weak. But because the very biological systems designed to protect us are now struggling to keep up. Society has evolved; the brain hasn't necessarily kept pace.

Your brain was not made to process a 24-hour news cycle, social media outrage, AI replacing entire industries, and the constant flood of information we face daily; it was built for running from predators, finding food, and keeping your tribe alive. The evolutionary wiring that kept us alive for millions of years now creates anxiety, stress, and cognitive overload in a world that changes faster than we can consciously adjust.

Disaster - real or imagined - throws a wrench into everything. A sudden trauma derails us not just in the physical world, but in the neural architecture of our minds. We spiral into worst-case scenarios. What if our medical tests come back with bad news? What if we lose the job, the relationship, the safety net? The vicious cycle of intrusive thoughts creates knots in our minds - triggers for fear, pain, and suffering that don't easily unravel. The brain, trying to conserve energy, defaults to the limbic system - the seat of fear, survival, and knee-jerk reactions - while the logical, rational frontal cortex takes a back seat. This isn't cognitive weakness per se. This is how the system was designed: **survive first, think later.**

George Miller's famous "Magical Number Seven" theory, backed by decades of neuroscience, shows that the human brain can only hold a limited number of pieces of information at once - even under calm conditions. Under stress? The field narrows even more. Instead of logical reasoning, the brain tunnels into survival pathways, hyper-focusing on perceived threats, real or imagined.

Picture standing on the edge of a cliff, heart pounding. The fear, the adrenaline, the tunnel vision - that's your limbic system hijacking the controls. Your frontal cortex - the seat of creativity, problem-solving, future planning - is nowhere to be found. Perhaps you were excited to embark on some mountain-scaling expedition, and your rational mind convinced you to try new things – but your ancient brain wasn't part of that dialogue.

Overreliance on the ancient limbic system was fine, and even preferable, when the biggest threat was a sabretooth tiger. It's a disaster when the threats are endless emails, layoffs, AI displacing industries, and the existential fear that we're falling behind in a world that's changing at warp speed.

The truth is, we are still evolving - but not fast enough. Evolution is no longer a passive process. In the age of AI, **we must evolve on purpose**. And the good news is, we can. Functional brain evolution - rewiring how we think, respond, adapt - doesn't have to take millennia. It can happen in real time, through conscious disruption and strategic training.

I designed a framework called **Ballistic Interruption** with precisely that intention at its' core: To intercept the automatic survival responses that no longer serve us, and to consciously reroute our brains toward resilience, logic, and creativity.

Ballistic Interruption is based on three critical components:

- **Biological Awareness/Interoception:** Knowing when your brain is sliding into survival mode.
- **Rationalization and Planning:** Activating the logical, reasoning centers of the frontal cortex.
- **Neuronal Visualization:** Rewiring your default responses by activating your vision center through repeatedly picturing new synaptic connections formed.

By training our brains to recognize and interrupt outdated survival patterns, we can build the kind of neural agility necessary to thrive in an AI-driven, constantly shifting world. We don't have to live stuck in old operating systems. We don't have to accept chronic stress, anxiety, or cognitive fatigue as normal. We have the ability - right now - to reformat and reboot our neural pathways for resilience, creativity, and mental clarity. But it's not just about surviving better. It's about **playing bigger**.

Most of us have been trained to think that success is about mastering habits: Wake up at 5 AM. Meditate for 20 minutes. Eat the right superfoods. Optimize every second of your day. Habits have their place. But they are not the foundation of real evolution. You can have perfect habits and still be mentally brittle - fragile in the face of real-world complexity, change, and uncertainty.

In a world where AI is redefining work, identity, and human capability at lightning speed, **neural agility** is the new survival skill. Creativity, adaptability, emotional regulation, strategic thinking - these are the traits that will define who thrives and who struggles. And they are all *trainable*. Neurons are wired to fire ballistically - once a pattern is triggered, it tends to run to completion. But with conscious training, we can create **new ballistic patterns** - ones that fire toward resilience, growth, and innovation rather than fear, paralysis, and retreat. We can choose to evolve. We can build brains that aren't just reactive but proactive. Brains that aren't just surviving but creating, connecting, expanding.

This isn't about becoming superhuman. It's about reclaiming the full range of human potential that evolution designed for us - and that modern life has dulled. Your brain is not a relic. It's not a fossil. It's a living, adapting, powerful engine of change - if you know how to use it.

In the chapters that follow, I'll show you 100 ways to actively future-proof your brain - practical, science-backed strategies to boost resilience, expand cognitive flexibility, enhance creativity, and protect yourself against the mental erosion of stress, fear, and overwhelm.

The future belongs to those who are ready to evolve - on purpose, by design. Let's get started.

Theme 1: The Basic Brain Operating System

Way 1: Fight Your Instincts to Keep Going!

At 24-years-old my doctor would tell me they suspected there was a hole in my baby's heart. A scenario that made no logical sense for a 24-year-old woman with zero history of drug usage or genetic abnormalities. After weeks of testing, the doctors would throw their hands up and admit the technician must've made a mistake. No apologies, no explanations. My relief was so pervasive that I could summon no vestiges of visceral rage at this nonchalant response. I recall thinking that I'd dodged a bullet and everything after that point would be easy and perfect. I was wrong.

At 6 months pregnant, I'd learn that Stage 4 cancer cells were found on my cervix and be given an ultimatum of having the affected part of my cervix cut out, risking miscarriage – or postponing surgery until post-birth, thereby risking my own life. Those 40 weeks of pregnancy I would find myself fighting not only for my own life on multiple fronts, but for the life of my child.

Many of my friends and family would constantly badger me with the painful logic *"you're young, you can always make another baby."* Those words seemed to admonish me; *"your baby is expendable, but you are not!"* My child was the most important part of my life and the fact that people I trusted were incapable of understanding the sheer cruelty of their words was agonizing. In the remaining months of my pregnancy, I recall isolating myself more from people who insisted their decisions for my health were more valid than my own. I decided to postpone surgery and wait until after my son was born to delve into cancer treatment. I didn't take time off work from my intense job at a global technology company. I pretended I was fine; I could handle everything. I kept going.

The human brain doesn't operate by Hallmark card timelines. Survival mode isn't an emotion; it's an instinct. It's messy. It's exhausting. It does not flip off the moment the immediate danger passes. Survival isn't clean. It isn't pretty. It doesn't always feel victorious. Survival often feels like emptiness - because the brain, still braced for impact, doesn't yet know the danger has passed. The half-life of "survival mode" is rarely predictable.

🧠 Future-Proof Your Brain Insight

When you survive something hard and don't feel "okay," it isn't weakness. It's your brain doing what it was built to do: detect threat, not celebrate survival. The instinct to "just keep going" is driven by your oldest operating systems: the **reptilian brain** and **limbic system**, which evolved for emergency responses - not for healing.

But here's the cost: Research from Dr. Bruce S. McEwen at the National Institutes of Health shows that prolonged stress physically reshapes your brain (McEwen, 2007):

- The **amygdala** (fear center) grows larger.
- The **hippocampus** (memory and emotional regulation) shrinks.
- The **prefrontal cortex** (planning, decision-making) weakens.

Without deliberate interruption, survival mode becomes your brain's new default - rewiring you for hypervigilance instead of growth.

To future-proof your brain, you must build a **"neural settlement."** A deliberate cognitive marker that tells your hippocampus:

"The threat has ended. It's time to rebuild."

This isn't about "positive thinking." It's about closing the survival loop so your brain can begin restoring emotional balance, memory formation, and higher executive function.

Think of it like climbing Mount Kilimanjaro: You don't sprint to the summit. You stop at settlement camps to recover, reset, and acclimate - or you collapse from the climb itself.

Your mind needs the same structured recovery. Until you consciously process and mark survival, you're trying to build a future on unstable, exhausted ground.

Your First Neural Settlement

- Mark survival as complete: *"I survived. That chapter is over."*
- Acknowledge the cost, without rushing to erase it.
- Give your nervous system permission to stand down.

You cannot architect resilience on a battlefield still wired for war. You have to call the ceasefire first; this is your winning move. This requires conscious and deliberate personal agency – it demands that you code the end of survival mode while your brain is still running rampant on fight or flight neurotransmitters.

"Just pause" is an oversimplified strategy. Pausing without recording your mental record is a waste of energy. Architecting neural resilience does not always require a pause – but it does require awareness and control; developing the ability to face current realities and interrupting your ancient operating system that was built millions of years ago.

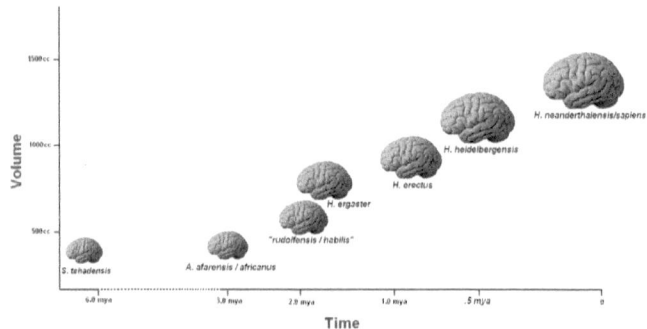

Image: Plotting the average size of the brain over time (Bolhuis, 2014)

In the same way that we can consciously pause our breathing and impact our heartrate accordingly, it is possible to learn to consciously control brain function through consistent interoception.

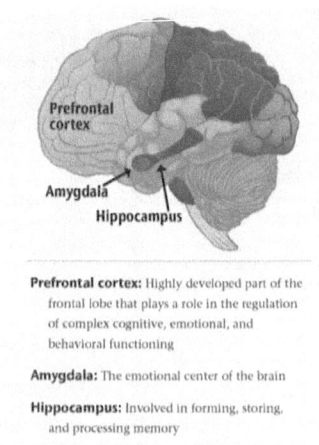

Image: Mindfulness and Brain Structure (Holzel et al., 2011)

Way 2: Challenge Your Brain's Prime Directive – Go Beyond "Stay Alive!"

Humans are the epitome of cognition evolution. Right? We like to think of ourselves as rational. Logical. Measured. But when I received the terrifying medical information about my child – all my lists and schedules and plans meant nothing. I went numb and felt detached from everything and everyone around me. Shock, anxiety, and panic took over my generally logical thinking. Not by choice. By design.

This is because, despite our smart phones, automated calendars, and unnecessarily complex coffee orders, our wiring hasn't progressed far from those early humans running from a tiger. When the brain perceives a threat, it does not send an activation signal to your "thinking brain" or frontal lobe. It bypasses it entirely. Instead, the limbic system - often called the "lizard brain or reptilian brain" - takes over. If you ever get the chance to dissect a brain (not always a fun experience), you will notice that the limbic system is closest to the brain stem and is the most insulated. There's a reason for this architectural design – you need your limbic system to survive. Think of it as your most basic line of code in an operating system that boots you into being alive.

This system, which includes the amygdala (emotion) and hippocampus (memory), is deeply ancient. Its' job is singular: protect the organism and perpetuate the continuation of the human species through procreation and survival. When my doctors shared crushing medical information with me about my baby's health and my health – living and surviving is ALL that mattered. I needed to LIVE and ensure my son LIVED – all the arguments from people around me insisting I be "rational" meant nothing. I have no regrets about my decision to choose my son. And here's the kicker, that reaction **wasn't a flaw of ancient modelling. It was a feature!**

In a life-threatening moment, your brain doesn't want nuance or philosophy. It wants survival. It wants to run, fight, or freeze - not weigh pros and cons. That's why even after the immediate crisis ended, even after my son was born – I kept going and acting like I needed to survive. To my limbic system the crisis wasn't over, and I was braced for another one to arrive. My body was running on cortisol and adrenaline, and it wasn't going to simply turn off that engine of chaos.

We often punish ourselves for how we "handled" stress. The not always kind internal dialogue berates us that we should've stayed calm, made better decisions, not overreacted. But the truth is: we weren't supposed to "handle it." We were supposed to survive it. Your brain's prime directive is not peace - it's protection. And it will choose your survival over your serenity every time. If you are "beyond" a trauma event and still frustrated with how you handled it in the moment, be kind to yourself. You're only human!

🧠 Future-Proof Your Brain Insight

Stop judging your body's stress response. Instead of asking *"Why am I like this?"*, ask *"What is my brain trying to protect me from?" "I trust that my brain is working hard to keep me safe, even if I don't understand its timing."*

You can future-proof your brain by removing shame and vilification from survival responses - and replacing it with curiosity and compassion. This unlocks awareness - and awareness begins the shift from survival into healing. But trusting in your brain potential is just the beginning – you're about to embark on a massive renovation of your entire neural circuitry in the pages ahead. That renovation is what will arm you to thrive in The Age of AI and successive future industrial revolutions. It will grant you the sustainable skill of constant neurological rewiring throughout your life cycle.

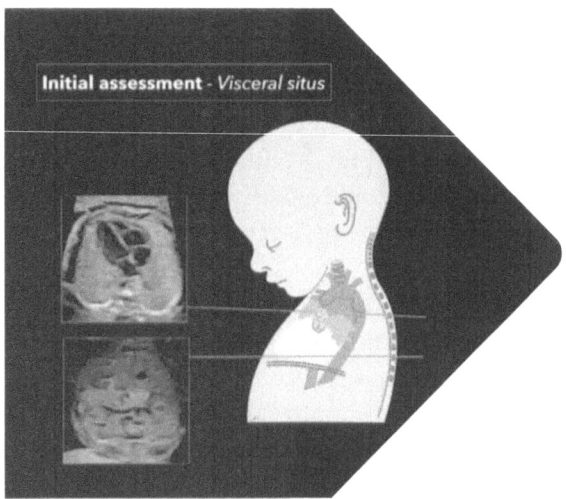

Image: Sample fetal echocardiogram – at 4 months pregnant when doctors believed there was a hole in my son's heart, they ordered one of these and I went straight into survival mode at Sick Kids Hospital in Toronto, Canada.

Way 3: Understand The Chemistry Behind Your Reactions!

When we see someone frozen in fear, we often interpret it as fragility. When we see someone fall apart after a breakup, a layoff, or a hospital stay, we think - *they're not coping*. But trauma has never been about coping, it's been about **chemistry**. Remember that intense trauma I shared of having to choose between my son's life and mine? Even after my son was born, and I had surgery to remove the cancerous cells and had verifiable evidence of clear tests… I still didn't feel "normal." The prognosis was good. Life was moving forward - at least on the surface. But inside my body, everything was still on fire. I had night terrors – I was constantly terrified something would happen to my son, that the cancer would come back, and he'd be left with no mother. No matter what my doctors told me, no matter how much research I did – that fear was pervasive. Forever known as the scientist and the one who was calm and collected – I was frustrated with my brain and its prolonged "weakness" as I perceived it.

Here's why: **my limbic system hadn't received the memo that the danger was over.** The amygdala, our brain's alarm bell, sits right above the brainstem - buried deep within the folds of the brain. It's so protected that it often survives trauma better than we do. That's its job: to remember danger, to sound the alarm - again and again - to keep us safe. Even if the threat is no longer real. Even if it's just a memory.

When trauma strikes, your brain forms rapid-fire connections between the event and everything around it: The smell of antiseptic. The sound of hospital monitors. The sentence the doctor said. The time of day. These fragments get lodged in memory - not as stories, but as triggers. Your brain doesn't organize trauma like a file. It replays it like a broadcast. Because repetition, to the brain, is the best way to stay alive next time.

"Remember this. Don't forget this. This almost killed you once."

That's why trauma feels like you're still there - still on the operating table, still in the fight. Because part of your brain is. And the more you try to "snap out of it," the more it resists. Because snapping out of it would feel like lowering your defenses. It's not weak. It's biology. And until we understand that - we'll keep mistaking trauma symptoms for personal failure. Or we tend to push towards the opposite side of the pendulum – encouraging people to completely shut down and not learn to thrive through and during chaos.

🧠 Future-Proof Your Brain Insight

The most powerful thing you can do in the aftermath of trauma is **name what's happening** as *biology*, not *character*. This one mental reframe can pull you out of shame and into science. *"This is my brain protecting me. This is chemical. And chemistry can change."*

To future-proof your brain, don't just try to out-think trauma. **Begin by telling your nervous system the truth:** You are safe now. You are not weak. You are wired to survive. And survival isn't failure - it's proof of resilience. The truth also includes factual understanding and a bit of revisiting BIO101 – framing what limbic function is; a hardcoded foundational system of survival – one ready to take control at the first sign of perceived threat or danger. Neuroscience knowledge doesn't only belong in labs and to PhDs and surgeons. Cognitive science learning should be as compulsory as knowing cardinal directions, basic math, or what H_2O stands for.

Looking back on those life-changing moments of fear I often what if GenAI conversations could've helped regulate my nervous system and been a neutral sounding board for all of my motherly angst.

Image: August 2023; Metro Hall, Toronto, Ontario
Sarah Baldeo, TED Talk: "The Neuroscience of Resilience"

Way 4: Understanding That Trauma is Generational AND Genetic

My parents divorced when I was 3 years old, and shortly thereafter both remarried. I gained two bonus parents – one Italian, one Russian. As an adult, I'm so grateful for my stepparents, but those early years were not characterized by easy shared custody. At age 13 I moved to live with my dad and stepmother, by 18 I left home. The adolescent experiences of living in one of the top gun violence neighbourhoods in Toronto – the times when 18-year-old me would choose between gas and groceries – these memories all made me who I am today. But before my own experiences, I know there are centuries of struggles my grandparents and great grand parents survived through constant immigration from so many different parts of the world. In every single generation of my family, including my own, immigration has been a constant. From China, Ireland, Spain, Jamaica, Guyana – and now for myself as a first generation Canadian.

There's a reason we call it *carrying the weight*. Because some of us are carrying stories we've never lived, pain we've never named, and fear that was passed down quietly, in blood and bone. This isn't poetry - it's biology. Welcome to **epigenetics**, the science of how our genes express themselves in response to our environment. You inherit your parents' eye colour. But you can also inherit the imprint of their stress. Evolution is typically classified as ontogenetic (individual) or phylogenetic (population wide) – but research has revealed that even if a child isn't subject to environmental stressors the weight of traumatic experiences can be inherited.

Here's how it works: trauma doesn't always change your DNA sequence, but it can change which genes get turned on or off. These changes, called **epigenetic marks**, can be passed to your children and grandchildren - affecting how their brains develop, how their nervous systems respond to stress, even how they metabolize fear.

Researchers have found this in children of Holocaust survivors, in descendants of those who lived through famine, war, slavery, and displacement. Even rats exposed to traumatic stimuli prenatal have passed stress sensitivities onto their offspring, despite in some cases the offspring being separated from mothers at birth (Weinstock, 2016). What we survive - or don't - ripples through time onto the next generation.

This means that your brain may be responding to danger that happened long before you were born. It may be holding the emotional residue of a grandmother who lost a child, a father who fled a civil war, or a mother who learned to stay silent to stay safe.

If your life feels like it's full of invisible battles, like you're always on edge, like joy doesn't come easily - it may not be your fault. It may be your inheritance. But here's the good news: **awareness changes biology too.** Epigenetics is not a life sentence. It's a wake-up call - a reminder that healing yourself may also free the generations that come after you. Rewiring not only your behaviour, but neural synapses in tandem, is an endeavour that yields a rich and valuable legacy.

🧠 Future-Proof Your Brain Insight

If you feel things that "don't make sense," stop asking what's wrong with you. Start by asking: *"What might I be carrying that isn't mine?" "I am the first in my line to interrupt this pattern. What was passed down ends with me."*

To future-proof your brain, make space for the possibility that some of your pain is ancient. And that healing it now **doesn't just transform your life - it transforms your lineage. You cannot currently select your baby's eye colour (but that ability is coming down the pike), but you can mitigate the effects of generational trauma, a proverbial vaccine against future stress and anxiety.**

> *"All children have to be deceived if they are to grow up without trauma."*
> - *Kazuo Ishiguro*

Way 5: Rewiring Your Instincts – from Limbic System to Frontal Cortex

Trauma teaches us that danger is constant. That control is a myth. That pain is permanent, and healing is a privilege other people get to feel. But that is a lie, not just emotionally - but neurologically. The human brain was not only built to survive; it was built to adapt. And that's our great (largely untapped) evolutionary secret.

Neuroscientists call this capacity **neuroplasticity** - the brain's ability to rewire itself in response to experience. New thoughts and behaviours can form new connections. New habits can strengthen new circuits. And with enough repetition, even deeply grooved trauma loops can weaken. Think, for example, of that childhood friend you have who outgrew being terrified of bees. Or how as a child the dark was terrifying but is now a welcome respite.

Yes, trauma can shrink the brain. Even the frustration with being unable/unwilling/unprepared to upskill on new AI tools can cause the brain to shrink in size. But attention, compassion, learning, movement, creativity - these can help grow it back. Not in the way it was before, rather in the way you need it be architected now.

Imagine my own story of becoming a mother. By the time my son was 2 years old I'd get separated, shortly thereafter divorced, and exit a marriage of extreme domestic violence. I'd embark on what would become 10 years of lone parenting with no financial support or shared custody. I had no way of knowing the challenging years awaiting me. While I realize now how deeply in survival mode I was mired, I also observe that I was laser-focused on moving forward. This is the crux of future-proofing your brain: You don't go back. You grow forward. Your trauma isn't undone. But it's no longer the only voice in the room.

The suggestion that we are at the mercy of our instincts does a disservice to the ability of the brain to grow, evolve, and change. Rewiring neural circuitry sounds like a monumental undertaking, but the entire system of neurons consists of ganglion and myelin sheaths. All of these structures were created for the purpose of evolving within one lifetime and basking in the benefits.

🧠 Future-Proof Your Brain Insight

Your brain is neuroplastic - not fixed. So don't wait until you "feel ready." You start changing your brain **by behaving like someone who is changing** - even in small, quiet ways.

"I can't go back to the person I was. But I can become someone new."

To future-proof your brain, lean into what builds new wiring:

- Movement
- Music
- New skills
- Safe relationships
- Rest
- Storytelling
- Curiosity

Neuroplasticity doesn't erase pain. It writes over it. And every time you choose growth - especially when fear tells you not to - you are choosing a new brain. Reactions ARE choices.

> *"Sons are the anchors of a mother's life"*
> *- Sophocles*

Way 6: Cognitive Flexibility & Being Your Own "Interior Designer"

When I was a single mother in my late twenties, money in downtown Toronto was tight on one income- but creativity was non-negotiable. I had a great job as a VP at a New York tech firm, but I still only had a single income in the most populated city in Canada. And I was determined never to return to the environment where I grew up. I wanted my child to grow up in safety, security, stability – and yes luxury. I was determined that he experience comfort and calm, as a means to learning foundational limbic control.

Every few months, I'd rearrange our home. Not always with new furniture or expensive upgrades, but with what we already had:
- I'd flip the living room layout
- Repurpose a dining chair as a side table
- Move art from one wall to another
- Sometimes I'd even paint something new myself
- Drape scarves over lamps for texture

To outsiders, it probably seemed like a quirky or obsessive ritual. But for me and my son, it was about more than aesthetics. It was a way to shift the emotional energy of the space. To make the familiar feel fresh. To remind ourselves that change was possible - even on a budget. What I didn't know at the time is that this habit was doing something profound inside my brain: It was building **cognitive flexibility.**

💬 Insight 1: Changing your physical environment helps change mental patterns.

When you rearrange your space, you activate the **parietal lobe** (spatial reasoning), **hippocampus** (memory re-mapping), and **prefrontal cortex** (planning and adaptation). This shift in spatial logic nudges your brain out of autopilot and into **active reprocessing** - which:
- Enhances creativity
- Builds neural connections
- Increases tolerance for uncertainty and change

Even moving one piece of furniture forces your brain to relearn the layout - and in doing so, reminds you: *I can think differently than I did yesterday.*

🧠 **Insight 2: Environmental novelty boosts mood, motivation, and self-efficacy.**

Research shows that small acts of **environmental novelty** can:
- Increase **dopamine** (motivation, reward)
- Improve **focus and memory consolidation**
- Support **emotional regulation** by reducing visual clutter and monotony.

You don't have to hire an exorbitantly expensive interior designer. You just need to shift your frame of reference – literally, with picture frames, and metaphorically.

🧠 **Future-Proof Your Brain Insight**

Your brain needs practice letting go - not just of old thoughts, but of old configurations.

"When I rearrange my space, I remind my brain that I'm still capable of change - no matter the budget."

To future-proof your brain, don't underestimate the power of movement within stillness. Sometimes, shifting your environment is the most tangible proof that you're evolving. And sometimes, a new perspective begins with simply switching the sofa and the bookshelf.

> *"If you want a golden rule that will fit everything, this is it: Have nothing in your houses that you do not know to be useful or believe to be beautiful."*
> \- *William Morris*

Way 7: This is the Sound of Silence

For those of us who had iPhones in the 2010s - Remember that U2 album that was automatically downloaded to everyone's iPhone in 2014? It would randomly play every time you connected your Bluetooth. As great as U2 is – listening to the sound of silence is JUST as important for your brain. In North America, we know the term "awkward silence," but in many cultures silence is not a space for discomfort. It's a place for respectful quiet and space to reflect, even in the company of another person. For much of the West, we are socially conditioned to think that chatter and filler is being a polite host or companion, rather than neurologically inefficient or a waste of mental energy. As a closet introvert, I love quiet – but years of being conditioned to be constantly entertaining and useful to those around me meant that quiet makes my nervous system sometimes go into overdrive. It used to feel like quiet was a space for uncertainty and emotional ambiguity, but part of upgrading my brain was unknotting those childhood traumas and building an ecosystem as an adult where love and acceptance isn't tied to being constantly "fun."

Meditation, a practice dating back to 1500 BCE, seems to be generally accepted as a positive activity – but it can feel like a good thing or an impossible thing. This practice is what most easily comes to mind when we think of silence. It's not always easy to try to emulate a Buddhist monk's practices in a modern urban space. And the reality is that most people do not have the time or the training to meditate effectively. But what sounds far more sustainable and possible, is finding five minutes for silence – no, not for "doom scrolling" on social media, but for actual silence. Active silence reduces brain inflammation, rids our brain of neurotoxins, and helps us to process data effectively. In theory it's an easy practice, and doesn't require a significant departure from, or destruction of, one's belief system.

It is important to distinguish between "active silence" and meditation. Let me clearly state that in neuroscience research these two practices are not equivalent. Actively listening to the sound of silence is not the same as going for a walk and listening to nature or listening to a white noise machine. Active silence is about depriving your brain of as much stimuli as possible – think about a detox for your senses. It's very hard to be still in silence and not reflect on everything you need to get done. It's far easier to try and use that quiet time "productively." But listening to the sound of silence has neural effects that can regenerate cells in your brain.

🧠 Future-Proof Your Brain Insight

- Active silence involves sitting for five to fifteen minutes listening to nothing at all: not meditation music or white noise, but silence.
- For those who struggle not to overthink, this process can also involve imagining (and not actually doing) a repetitive task, which can regulate the nervous system.
- We can expand this practice as we become accustomed to it.
- Researchers at Duke University (Kirste, 2013) discovered that listening to the sound of silence kickstarted neuronal regeneration in the hippocampus – the part of your brain that controls memory. Listening to silence caused new brain cells to grow!

> *"In silence, we often find solutions to problems we can't solve in any other way."*
> *- Eknath Easwaran*

Way 8: Interrupting Those Automatic Negative Thoughts

Ever notice how much easier it is to recollect negative experiences than positive ones? Your brain is not trying to sabotage you. It's trying to protect you - but it's stuck in an old operating system. Picture for a moment the most atrocious experience you've ever had. Now pause. Then think about the most wonderfully fulfilling experience you've ever had. You might notice that it's far easier to recall the negative experience than the positive. Perhaps you believe that's because I asked you to think about the negative first, but even if I reversed this exercise, you'd still have a far easier time of recalling the horrendous memory.

That system is called **negativity bias**: the tendency for your brain to prioritize, encode, and replay negative experiences more intensely than positive ones. You're not broken. You've *evolved* that way. In prehistoric times, remembering which berries made you sick or which path led to the lion's den could mean the difference between life and death. So, your brain became exceptionally good at tracking threats. That system kept you alive, but now it is what makes you overanalyze one awkward comment, agonize over a text reply, or replay that one meeting that went sideways - instead of celebrating what went right.

💭 **Insight 1: Automatic Negative Thoughts (ANTs) are neural shortcuts - not truths.**

ANTs are mental reflexes - quick, reactive interpretations like:
- "I'm not good enough."
- "They probably don't like me."
- "This will never work."

These thoughts come from your **amygdala** and **default mode network**, both of which are tuned for **pattern recognition** and **self-preservation** - but not nuance. Over time, if left unchecked, ANTs create **entrenched neural grooves** that shape how you see the world - and yourself. Imagine well-worn ruts in the road of your mind – it's often just easier to take the roads more travelled.

Interrupt, Label, Rewire

The first step is not to deny the thought - it's to catch it.
- Notice it ("There's that fear of failure again.")
- Label it as an evolutionary echo, not an accurate reflection

- Replace it with a curious, constructive alternative

You're not arguing with the thought. You're refusing to be automated by it. This activates the prefrontal cortex, the part of your brain responsible for rational decision-making, emotional regulation, and perspective-taking. This isn't toxic positivity. It's pattern disruption - and pattern disruption is how you build a different brain.

🧠 Future-Proof Your Brain Insight

Your brain remembers pain faster than joy - because it evolved to survive, not to thrive. *"When I catch a negative thought, I interrupt an outdated survival story."* To future-proof your brain, learn to pause the reflex. Insert space between stimulus and response. Not to silence your past - but to **author your future**.

> *"All negativity is caused by an accumulation of psychological time and denial of the present."*
> - *Eckhart Tolle*

Way 9: The Novelty Pill

Your brain loves efficiency - but it grows in challenge, which occasionally is not the most efficient pathway. And nothing challenges it more gently and powerfully than novelty. Novelty keeps you flexible, focused, and future-ready. Trying something new once a week doesn't just expand your experience; it literally reshapes your neural wiring. The things that feel small - taking a different route to work, ordering the dish you can't pronounce, reading a genre you've never explored - these **micro-novelties** activate major cognitive shifts when they become an entrenched element of your everyday life.

We also see this in GenAI over-reliance. Introducing novelty into prompts that we engineer increases brain function and activity in users and decreases AI reliance. On average, conversational AI tools provide higher quality responses when they are dialogued with at least 2-3 times. That requires intense self-awareness and the confidence to push back against AI, even as you are learning to use it.

Insight: Novelty activates dopamine and increases neuroplasticity.

When you encounter something new, your brain releases **dopamine** - not just for pleasure, but to encourage **exploration** and **learning**. That rush of newness fires up the **hippocampus (memory)** and **prefrontal cortex (planning + decision-making)**, improving:
- Attention span
- Emotional regulation
- Memory consolidation
- Problem-solving ability

Novelty isn't entertainment. It's **mental resistance training**.

The Practice: Schedule a small novelty every 7 days
- Walk around a new neighborhood
- Try a new spice or recipe
- Watch a foreign film
- Use your non-dominant hand for something routine
- Rearrange your bookshelf

When you do this **weekly**, your brain gets regular doses of:
- Pattern interruption

- Mild uncertainty
- Rewarded curiosity

All of which fortify cognitive flexibility, making you more adaptable to change - even the kinds you don't choose. And here's the real beauty: You don't have to like the new thing for your brain to benefit from it. In fact, sometimes it's best to try things you are sure you won't like. You just must do it on purpose.

🧠 Future-Proof Your Brain Insight

Repetition makes you efficient. But novelty makes you antifragile.

"When I expose my brain to something unfamiliar, I train it to stay brave."

To future-proof your brain, don't let comfort become calcification. Seek out surprise. Reward the unfamiliar. Train your mind to expect and embrace change - and meet it with curiosity, not resistance.

> *"I love scandals about other people, but scandals about myself don't interest me. They have not got the charm of novelty."*
> - *Oscar Wilde*

Way 10: Active Problem Solving – no AI-assistance!

In the Age of AI, we're outsourcing more than tasks - we're beginning to outsource thinking itself. Actively solving problems independently engages your critical thinking and frontal cortex. Search engines finish our sentences. Generative AI (GenAI) tools complete our ideas. Predictive text finishes our text messages. Recommendation algorithms tell us what to watch, read, and think about next. But here's the warning: easy is not always adaptive. And every time we let a tool think for us, we lose a tiny bit of our own neural stamina. This doesn't mean "stop using AI," rather it means be aware of dependency on it.

A few months ago, I found myself using GenAI to draft sample homework algebra tests for my teenager who is in Grade 9. I was admittedly too exhausted from work to create sample tests on distributive property, factoring, differences of squares, and linear equations (but I was happy to discover I hadn't forgotten any of that math in the 20+ years since high school!).

I discovered something, outsourcing this task to AI was not helping me to help my son with studying, in fact it was making my brain even lazier. I was often tempted to skip over validating his answers and just look at the answer keys instead of following each step of his mathematical proofs. I quickly admonished myself – here I was, someone who'd taken all her high school Advanced Placement tests in Math and landed scores of 5's on all of them.... looking up answer keys. It didn't exactly put me in the position to lecture my son on not using ChatGPT for his homework.

🧠 **Insight: Active problem-solving strengthens cognitive flexibility, decision-making, and executive control.**

Active problem-solving involves:
- Identifying the problem
- Generating multiple solutions
- Testing, revising, iterating
- Tolerating uncertainty
- Making decisions without perfect data

It's the work your **prefrontal cortex** was built to do.

By contrast, **passive problem-solving** is when we:
- Wait for external resolution
- Hope the issue disappears

- Avoid conflict or decisions
- Let others step in
- Or rely solely on tools like AI to "figure it out for us."

Over time, passivity erodes the neural circuits that make us resilient, creative, and self-trusting. You don't have to avoid AI - but you must know when to think before you prompt. Create and iterate your OWN ideas and then dialogue with conversational AI tools. Because if you don't practice making decisions without help, your brain won't remember how, and it will be completely disincentivized from using your laudable critical-thinking skills. This cognitive scaffolding is identical to why we must teach out children to do things for themselves in order to grow into well-adapted contributing members of society.

Future-Proof Your Brain Insight

Passive minds get predictable. But active minds get powerful.

"Every time I solve a problem without outsourcing it, I sharpen my capacity for the next one."

To future-proof your brain, schedule time each week to work through something without assistance - no AI, no search engine, no default delegation. Because your brain doesn't grow by watching answers appear. It grows by building the path there yourself.

> **"The rise of powerful AI will either be the best or the worst thing ever to happen to humanity. We do not yet know which."**
> *- Stephen Hawking*

Way 11: Interrupting Panic with Logic – Rewiring your Survival Response

You can palpably feel it when it starts. Panic. Anxiety. Fear. Are you in control? You want to be, but you're not quite sure. The heart races. Your chest tightens. Your thoughts accelerate like dominoes falling in slow motion - fast, unstoppable, inevitable. Wait a second…you're officially out of the driver's seat now. A memory, a headline, a text, a noise. And yes, even night terrors. And suddenly, you're launched - into fear, panic, or rage.

We often think these reactions are beyond our control. And sometimes, in the moment, they feel like they are. But neuroscience gives us a radical fact!: we can interrupt them. Mid-flight. The term is called **ballistic interruption** - a way of breaking a mental pattern even as it's happening, not after it's over. It's a practice I devised in my thesis research work when I was researching how the brain can handle multi-tasking. When neurons fire, they trigger a chain of activity. Trauma wires the brain for speed and protection. The brain fires fast, like a missile. That's the "ballistic" part. But interruption? That's the magic. That's where your rational brain - your **prefrontal cortex** - can step in and create a new pattern on purpose.

You can speak aloud. You can move. You can breathe. You can ask one clarifying question. These actions force the brain to switch networks - from the reflexive limbic system to the rational, regulating parts of your mind. It's like flicking a breaker switch to reroute electricity. You don't erase the feeling - you simply divert the power. Ballistic interruption isn't about perfection or control. It's about micro-mastery. Just one decision. Just one redirected impulse. Just once. And if you do it once, your brain starts recording it as a new option. Over time, the missile loses its trajectory. The pattern weakens. And your brain learns something new: you don't have to go all the way down with every spiral.

I'm frequently asked about being a 6x Founder with 2 exits. For the non-finance readers that means I've sold two of my companies. The most successful company I built was created and scaled during the most chaotic time of my life – while I was concurrently battling domestic violence, divorce, and financial strain.

🧠 Future-Proof Your Brain Insight

When you feel the launch sequence start, you don't have to stop it. You just have to *interrupt it* - even for a second.

"This is a signal. I don't have to follow it. I can choose a new route."

Future-proofing your brain starts with practicing one act of **rational disruption** during emotional escalation. That's not just healing. That's rewiring, that's staying in the driver's seat and using the newest part of your brain – moving beyond your emotion center and upgrading your reaction to include logic.

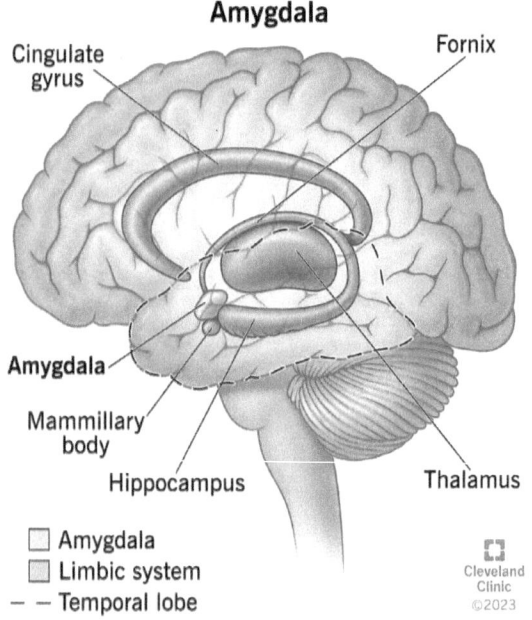

Source: The Cleveland Clinic, April 2023

Way 12: Be Willing to Re-align Your Memories

How often do you look back on your memories and reflect that perhaps you are recollecting the facts in a skewed or biased manner? How often do you question yourself? It's very easy to see how delusional others may be, but do you challenge your own bias? I remember my son as the perfect little toddler, always listening to me and never throwing tantrums – but ask my best friend or his godmother? They'll tell you all about the time he had a fit at 2 years old because I refused to buy him a toy. Or when he was 5 years old and refused to go to his karate class by sitting in the middle of a busy road.

We are primed to experiences in ways that suit us. Why? Because sometimes we need to build a narrative that makes us feel in control and safe. Remember your childhood sweetheart – they were perfect for you right? Or maybe you recall that the bedroom you grew up in as a child was always immaculately clean – or perhaps you were the perfect child and never challenged your parents, at least from your perspective.

We hold all our memories as a reality from which we navigate the world. Memory holds the breadcrumbs for the Hansel and Gretel in each of us to find our way back "home," and home is just another word for a place that feels safe. When life changes or we're confronted with some kind of untruth, our brains go into low-level shock, where we relive the past and recreate our future. Remember that axiom: "the best predictor of future performance is past behaviour." The unpredictability of memory function is precisely why eye-witness testimony is often considered unreliable in court proceedings.

🧠 **Insight: We build new memories more than we focus on editing old ones.**

- Our memory system builds foundational premises over time, which govern what I call our truth system, the long-term 'facts" that we use as a baseline from which to navigate our lives.
- When we experience a life change such as a trauma, our brains must recode these memories to make things true again – a painful exercise for the human brain, which prefers stability and efficiency. Think about what happens to a person when, for example, they discover infidelity in their marriage.

- The advantage of neuroplasticity means that this is an opportunity to utilize the brain productively, rather than focusing solely on preserving neural resources by refusing to change.

🧠 Future-Proof Your Brain Insight

Your brain isn't trying to hurt you - it's trying to **keep the world consistent**. But old memories aren't always the truth. They're just the *first draft of safety.*

"I can let my brain learn something new - even if it challenges what I thought was true."

To future-proof your brain, allow your truth system to evolve. Your memory center, the hippocampus, doesn't love to rewrite memories, it's painful and neurologically taxing – but it is often necessary to build a stronger and healthier brain. **Memories are not walls. They are doors.** And you have the power to decide which ones still belong in your home, as well as the ability to admit that sometimes in the moment that something is happening to you, your brain perhaps records it in a more palatable way than what reality offers.

> **"Sometimes you will never know the value of a moment, until it becomes a memory."**
> **- Dr. Seuss**

Way 13: Challenging The Fairytale Archetypes

Hansel and Gretel followed breadcrumbs. Cinderella waited for rescue from a heinous stepmother. Snow White stayed asleep until someone kissed her. Rapunzel let down her golden hair, and Princess Jasmine waited for her Prince. We grow up on stories like these - tales of obedience, reward, rescue, destiny. But here's what no one tells you: these early childhood stories take up permanent residence in your brain. Not as fiction - but as expectation. And they shape what you believe is possible.

From a young age, fairy tales create archetypes that become internal blueprints:
- Be good, and good things happen.
- Follow the path, and you'll find happiness.
- Stay silent, and someone will save you.

These aren't just bedtime stories. They're neural pathways - mapped into your social and survival systems. Why? Because fitting in used to be a matter of life and death. Thousands of years ago, stepping outside the group meant losing food, safety, or protection. So, our brains evolved to reward compliance - with dopamine. When you do what's expected, the brain lights up. You're safe. You belong. You live. But in modern life? Compliance doesn't always keep you alive. Sometimes, it keeps you *small*.

We're told to pursue the house, the title, the marriage, the smile. But following the rules can also mean lowering your expectations. You accept less. You need less. You forget what you wanted in the first place. And your brain, which thrives on familiarity, reinforces this path. Until one day you realize: You've been living inside a story you didn't write.

This is why reprogramming your brain starts with rewriting your narrative. Before you try neural training, or ballistic interruption, or any future-proofing techniques - You must ask: Whose story am I in? And do I want to stay here?

🧠 Future-Proof Your Brain Insight

Your brain rewards you for staying in line. But healing - and evolving - often means **leaving the script entirely.**

"I can unlearn the story I inherited - and write the one I need."

To future-proof your brain, you have to disappoint the old archetype. This isn't rebellion for its own sake. It's biology growing beyond mythology. You are not a fairy tale. You are a living system, capable of authorship. Instead of waiting to be saved, perhaps you became your own saviour, maybe you disappoint your parents, maybe you don't retire at 65 and maybe you don't ever fulfill any dreams other than your own.

Spoiler Alert: We don't have to go through trauma to be worthy OR find happiness. Fairytales teach us this trope of trauma being a necessary part of any hero's journey, but survival is not a competition. This is not the "Hunger Games" and you are not Katniss.

Vintage Engraving from the Brothers Grimm "Hansel & Gretal"

Way 14: Don't Stop Evolving

There's a moment in every healing journey when you realize: knowing isn't enough. You can understand trauma. You can name it. You can explain it to others. But eventually, your brain doesn't need more information. It needs transformation. That's the invitation of evolution. And it doesn't belong only to Darwin or biology textbooks - it belongs to *you*.

There are **three key steps** to training your brain to evolve - not just recover (this is the "Ballistic Interruption" I touched on in Way 11):

1. Biological Awareness

The first is what you've been doing so far: learning how your brain works. When you name the limbic system, identify triggers, understand stress hormones, you begin to shift from *victim of experience* to *witness of process*. This awareness is foundational. But awareness alone is passive. The next step is what makes it active.

2. Planning and Rationalization

Once you recognize a pattern, you can *plan around it*. You can begin to say:

"*This is the old loop. What's one small way to interrupt it today?*" This isn't about erasing your wiring - it's about building a **bridge** to the new. And like any bridge, it needs repeated steps across it to become strong. Which leads to the final step - the one that turns thinking into architecture.

3. Neuronal Visualization

This isn't science fiction. It's your brain's natural superpower. When you imagine yourself acting differently, or even see yourself building new thought pathways, you're activating real synaptic change. The brain doesn't distinguish well between imagined action and lived action - it fires similar patterns. This means you can rehearse who you want to become, and the brain starts treating that imagined self as a viable path forward.

In a world that keeps throwing us change, uncertainty, and loss, the question isn't "Can I stay the same?" It's: How fast can I adapt - and what kind of brain do I want to build? Because the brain's evolutionary design stopped adapting to modern life a long time ago. It's up to us to consciously and deliberately pick up where nature left off.

🧠 Future-Proof Your Brain Insight

Your brain is not done growing. But it needs your permission - and your participation.

"I am not here to be the same. I'm here to evolve on purpose."

To future-proof your brain, don't wait for breakdown to become breakthrough. Practice mental evolution daily - through learning, unlearning imagining, planning, and acting - even in small, simple ways. Each time you do, you're not just surviving this moment. You're designing the next one.

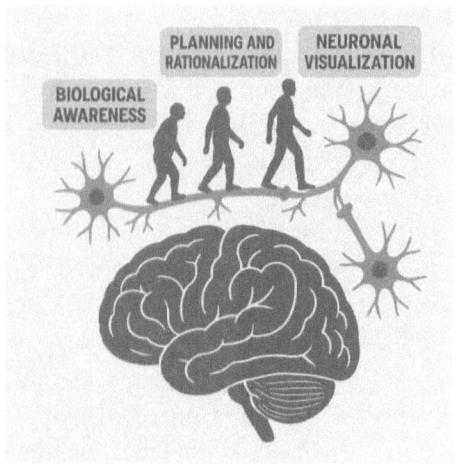

**Image: The Steps of "Ballistic Interruption" for a neural trigger – How to stop your brain from staying stuck in panic.
Source: ID Quotient Research Labs, 2023**

Way 15: CHOOSING Change vs. Letting it Happen

Your day doesn't need to start with a crisis for your brain to act like it's under attack. A terse email, a honking car, a client who says, "Let's circle back." Your amygdala really doesn't know the difference between being chased by a lion and being cc'd without warning. To your brain, change is a threat. And it will default to panic, irritation, or shutdown - even if the change is harmless. When you know how to navigate what's coming you feel safe – but change is full of the nebulous, the ambiguous, and the potentially harmful.

So how do we train the brain to handle pressure better?

By practicing change when it doesn't matter.

This might sound silly, but it's deeply powerful:

- Wear mismatched socks.
- Rearrange your kitchen drawers.
- Take a new route to a meeting.
- Switch your dominant hand when brushing your teeth.

These aren't just quirky habits - they're **low-stakes neural stress tests**. Each time you *intentionally* break a familiar pattern; your brain lights up with new neural activity. It becomes less dependent on certainty and more comfortable with improvisation.

This is called **cognitive flexibility**, and it's one of the key pillars of brain resilience. The more you rehearse *"this feels weird, but I'm okay"*, the less likely your brain is to freak out during real, high-intensity change. In a world of AI upheavals, climate shifts, economic tension, and algorithmic anxiety - **calm is a competitive advantage**. But calm isn't a default setting. It's a trained skill. And the training can start with small things…like socks. Yes! Socks!

Here's a simple analogy – you don't wait to start training at boxing on the day you expect a fight. Instead, you condition your muscles and mind to build ingrained habits. Your brain requires cognitive training and conditioning in order to effectively hijack a fight or flight response that is millions of years old.

💭 Future-Proof Your Brain Insight

The best way to prepare your brain for big change is to **invite tiny changes into your daily routine** - without consequence.

"If I can change something small on purpose, I can handle something big when it happens." To future-proof your brain, don't wait for chaos. **Create controlled surprises.** You're not being difficult. You're being *adaptable on purpose.*

The Mismatch Socks Experiment: Try It

Way 16: Embrace the Weird, the Bizarre, the Discomfortable!

I love my French press coffee. Every morning, I have my routine of grinding the beans, letting the coffee sit in the press, breaking the "crust" after a timed two minutes, and eating precisely 3 mini almond biscotti. It feels cozy. And if I'm honest? It feels great to be in control and to know exactly what to do without having to weigh every decision. But if I want to make my brain work more effectively, it is incumbent upon me to build some discomfort into my life. Although this may sound like self-inflicted torment, disruption in our lives creates more ballistic interruption, constantly reducing our reliance on that old neural hardware, the reptilian brain. We must disrupt our beloved routines, especially our most comfortable ones, if we want our brains to remain nimble and healthy. The human brain went through its most intense periods of evolution during moments in history when we were nomads exploring the unknown world and sometimes fatal environment.

🧠 **Insight: Muscles are not built through continuing to lift the same weights. The brain is not built through a refusal to explore.**
- Science tells us that efficiency is the way the brain was designed to utilize fewer resources so it can process information faster. It sounds counterintuitive, but the brain was designed to be lazy, and in many ways, GenAI tools only help with this laziness.
- But if you can challenge yourself through discomfort, your brain will not be able to go into default mode. For example, if you have your utensils in a particular drawer in your kitchen, and you can move them once a week, move them…or perhaps to keep your family calm, once a month.
- Avoid using GenAI tools to solve problems - instead use them for research purposes. Think about how your brain creates a map of information when you research and review data and information - even information that is divergent from your eventual conclusion. When ChatGPT just gives you answers, you're missing all critical thinking.
- The perfect analogy is teaching a child to search for answers, rather than simply giving them the solutions.

"Do one thing every day that scares you."
- Eleanor Roosevelt

Way 17: Consciously Switch Environments

When my son started kindergarten, I felt a deep urgency. Not just to educate him - but to *expand* him. We already lived in a massive, multicultural city, but that wasn't enough. He was already used to the CN Tower, skyscraper, the noises of a bustling metropolis. To him Toronto was the norm.

I wanted him to see the souks of Tangier full of brilliant colours, the rhythms of Lagos, the temples of Phuket, the coastlines of Cozumel. Admittedly, part of this was for me - a reclaiming of childhood dreams my family couldn't afford. But more than that, it was intentional: I wanted to build a future-proofed brain - one that didn't freeze at the edge of the unfamiliar. Full disclosure – I was making up for the experience I never got to have at his age, veritably reliving my own childhood (as parents are wont to do).

🧠 **Insight: The brain grows fastest when exposed to new environments that contrast existing norms.**

Environmental switching - from city to nature, from structured to chaotic, from familiar to foreign - activates a wide network of brain regions, including:
- The **hippocampus** (spatial mapping and memory)
- The **parietal cortex** (sensory integration)
- The **default mode network** (introspection and adaptation)

This sensory contrast increases **neural plasticity**, improves **pattern recognition**, and sharpens **emotional regulation** - all foundational traits in a world that demands constant change.

City vs. Nature: Both Are Essential
- Urban environments stimulate the prefrontal cortex and reward systems. They promote speed, efficiency, and social calibration.
- **Natural environments** downregulate the nervous system, enhance **theta wave production**, and boost **creative cognition** by giving the brain a break from overstimulation.

Moving between these two environments regularly teaches the brain to flex. To pause and accelerate. To observe and adapt. And this doesn't have to mean global travel. You can drive from downtown to a trailhead. Or go from a café to your balcony. **The shift is the signal. We must** continuously challenge the evolutionary mechanism that drives us to seek "familiar caves."

🧠 Future-Proof Your Brain Insight

The brain you build indoors can stagnate. But the brain that travels - even a few miles - **learns how to reboot.**

"When I choose new surroundings, I choose to grow out of my defaults."

To future-proof your brain, don't just take breaks. **Take shifts.** Go where your senses recalibrate. Let your environment teach your nervous system how to change - and still feel safe doing it.

> *"One's destination is never a place, but a new way of seeing things."*
> *- Henry Miller*

Way 18: Get Off Screens!

We've long been told to limit screen time for kids - because their brains are still developing. But here's the irony: So are ours. Neuroplasticity doesn't stop at 18. Your brain is constantly rewiring - and what you feed it, how you use it, what you neglect - it all matters. Screens aren't inherently evil. But they are inherently overstimulating and neurologically narrow. And when screens become our default environment, we don't just lose time - we lose motor engagement, sensory complexity, and attention bandwidth.

Have you ever tried to have a coherent conversation with your teenager after they've been staring at their phone for hours? It's like speaking to a zombie…. or worse, the wall! At least zombies react to noise and movement. After they've been peeled away from those little screens for a few hours, suddenly they become human again and can communicate beyond monosyllabic replies.

🧠 **Insight 1: Screens overstimulate the visual cortex - but under-activate the motor and sensory systems. Get off that TINY keyboard!**

Typing uses minimal fine motor movement and repetitive action. Compare that to writing by hand, which engages:
- The **motor cortex** (hand precision and movement)
- The **cerebellum** (coordination and timing)
- The **visual-spatial system** (layout, pressure, visual feedback)

Writing, sketching, and manipulating objects in 3D light up more of your brain than tapping a screen ever can.

🧠 **Insight 2: Prolonged screen exposure is linked to reduced attention span, emotional regulation, and creativity.**
- Increased screen time leads to **thinner grey matter** in regions tied to focus and language in both children and adults.
- We all think we can perfectly multitask while on a phone, but multitasking with screens lowers **cognitive endurance** and increases **mental fatigue.**
- Excessive scrolling reinforces the brain's **dopamine reward loop**, making it harder to feel satisfaction from real-world, slow-reward experiences like reading, crafting, or conversation. Screens hack your brain's reward system. Those constant dopamine spikes of switching to

new forms of entertainment fuel adrenaline (and addiction) - but don't help you build real skills to navigate complexity or ambiguity.

🧠 **Insight 3: Time off-screen enhances memory, mood, and vision itself.** Just 60 minutes off-screen and in natural light can:
- Improve **depth perception and eye coordination**
- Reduce symptoms of anxiety and digital fatigue
- Reset your **circadian rhythm**, improving sleep and cognitive function

Your brain and body evolved for movement, sunlight, and multi-sensory engagement - not blue light and passive consumption. This isn't about quitting technology. It's about **breaking the cycle of dependency**. Even 15 minutes off-screen - journaling, walking, listening to the wind - helps rebuild your **default mode network** and give your prefrontal cortex a break from micro-decisions and fractured attention.

🧠 **Future-Proof Your Brain Insight**

Screens are tools. But your brain is an ecosystem. And ecosystems can't thrive in artificial light alone. *"When I step away from the screen, I step toward more of myself."* To future-proof your brain, don't just "unplug" occasionally. Create **intentional time in the physical world** - handwriting, movement, eye contact, and real texture. Because your brain wasn't built for endless tabs. It was built for touch, light, story, and silence.

Way 19: Break The Toxic Relationship with Endurance

During five years of custody battles and juggling lone parenthood after divorce – my copying mechanism was to accept and internalize that level of chaos as my baseline of normalcy. I trained my brain at that "weight-class" to thrive. Anyone who has been through a brutal divorce, a protracted lawsuit, or years inside a toxic job knows something about resilience that no podcast ever taught them: sometimes, it's not about bouncing back - it's about not breaking completely. Ballistic Interruption without neural resilience would be a short-lived solution. How do you build control in line with resilience?

In pop culture, we glorify habits:
- 5 a.m. wakeups.
- Green smoothies.
- Journaling beside Himalayan salt lamps.

But real life? It builds habits too. Habits of stress. Habits of vigilance. Habits of survival. When you endure long-term strain - be it emotional, financial, or relational - your brain adapts. It does not just "take it." It *records it* - and builds neural pathways to sustain that burden. Which is why, after the trauma ends, you do not necessarily feel free. You feel numb. Or suspicious. Or exhausted, even on calm days. This is your brain stuck in a groove - a habit of hardship. But here's the miracle: you can build resilience actively, not just reactively.

Resilience isn't just surviving dreadful things. It is the *ability to stay functional* while experiencing them - and to recover capacity when they're over. One way to do this? Consciously interrupt the narrative that things will always feel this hard. Stop the internal catastrophizing that says, "this will never end."

That can mean something as simple as:
- Staring at art while listening to ambient music.
- Sitting in silence on purpose.
- Writing down a thought you wish you could believe - and reading it until it starts to feel real.

These aren't affirmations. They're neurological shocks of calm.

Your brain has been conditioned to expect strain. But you can teach it what safety feels like, too. Resilience isn't passive. It's a skill. And like any skill, it grows with use - and rewires the brain as it does. "Brain gains" are weightlifting exercises for the brain – to the point that neurological vigilance begins to exist outside of adrenaline and disaster.

🧠 Future-Proof Your Brain Insight

Surviving hard things makes you strong. But learning to experience ease without suspicion? That's next-level resilience.

"My brain remembers pain - but I can teach it peace."

To future-proof your brain, practice calm with the same repetition you once practiced coping. Not as denial - but as preparation. Because your nervous system deserves more than endurance. It deserves evolution. Your very brain cells, neurons, were architected with efficiency in mind. Nerve fibers are surrounded by a superconductor highway – a sheath of protein that thickens with behavioural repetition. Think of it as electrical insulation where repetition makes the electrical pulse move faster with each cycle. This is precisely how to further invest in evolving your neurology.

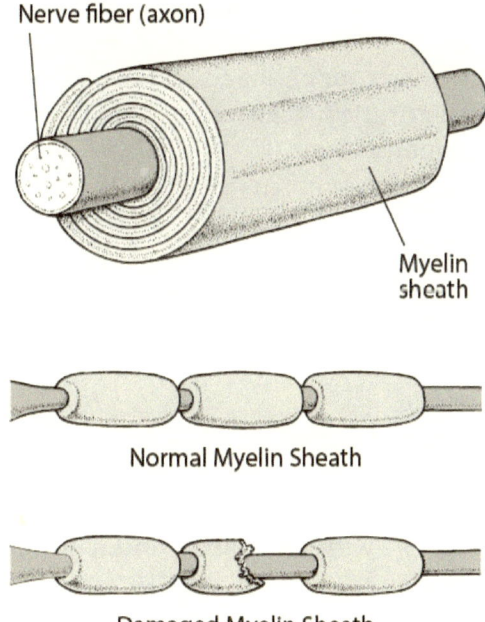

Image: The myelin sheath, discovered by Anatomist Rudolf Virchow

WEEKEND BRAIN RESET: A 24-Hour Screen Detox for Cognitive Recharge

You don't need to throw away your phone to reset your brain. You just need a short, deliberate pause - one that allows your nervous system to **re-sync with the physical world**.

Here's a 24-hour reset protocol you can repeat once a week, once a month, or whenever your brain starts to feel "fuzzy."

1. Off by 9 PM Friday

- Power down all non-essential devices (yes, especially the "background" TV or constant news cycle of doom).
- Charge your phone **outside the bedroom – read this as outside of your reach**.
- Replace your night scroll with a physical book or journaling session – a tangible writing or reading object that you can engage with.

2. Morning Light, Not Blue Light

- Upon waking, **get 10 minutes of outdoor natural light** (even if it's overcast).
- Delay checking your phone for at least **60 minutes.** If you can go back to a physical alarm clock instead of using your phone, this is even better for your cognitive health.
- Drink water, breathe, stretch, or walk - let your brain wake up without demands.

3. Use Your Hands

- Handwrite a to-do list, sketch something, cook from scratch, or do a tactile hobby.
- This activates **motor, memory, and emotion centers** in a way screens can't replicate.

4. Anchor in the Physical

- Spend 2+ hours engaged with your body or environment: walking, cleaning, hiking, gardening, crafting.
- Bonus: do it **without audio input** to reset your auditory processing bandwidth.

5. Converse With No Devices Nearby

- Share a meal, go for a walk, or sit in conversation without screens between you. Challenge yourself to avoid mentioning topics that saw on social media or googling something to prove that you're right. Rebuild human interaction patterns that are not device-based.

- Eye contact, vocal intonation, and micro-expressions **stimulate empathy circuits** and strengthen social intelligence.

6. Evening Reflection, Not Re-entry
- Don't "re-plug" into social media or news at bedtime.
- Reflect on how the detox felt. What did you notice? What came back online in you?

Way 20: Neural-Trigger Mapping

Your brain is not a blank slate. It's a **pattern-matching machine** built from lived experiences, inherited wiring, and subconscious bias. What soothes one person may completely unravel another - and unless you've taken the time to notice your personal blueprint, you'll keep assuming everyone's mind operates like yours. A fatal mistake and one that's caused far more arguments that can be counted.

You know the series of "love languages"? Well, the brain also has its own preferred modalities – and each brain is unique. Let me give you a personal example. About a year ago I got remarried after being sure I never would. What calms my husband? Waking up to *The Financial Times* blaring as he scrolls headlines and parses economic data while making a delightfully colourful shakshuka.

What calms me? Solving problems the moment they arise - clearing mental clutter by taking action. And I tend to do this first thing in the morning! I love it – I feel like I'm seizing the day – carpe diem! But for him, my need to resolve everything immediately feels like pressure. And for me, his news audio-track feels like waking up inside a stock exchange. Same household. Same morning. Two completely different neural triggers - one activating stress, one reducing it – and for each of us the opposite sources.

🌑 **Insight: Neural-trigger mapping is the act of identifying your unique inputs for stress and calm, so you can navigate your world more intentionally.**

Every brain has a **sensory and emotional pattern**:
- Some people are calmed by noise, others overstimulated by it.
- Some people regulate through logic, others through emotional connection.
- Some need resolution. Others need space.

These patterns are driven by your:
- **amygdala** (threat detection)
- **insula** (internal body awareness)
- **prefrontal cortex** (interpretation + regulation)

The more you understand **what activates or soothes these systems**, the better you become at:
- Managing your own nervous system

- Communicating with others
- Choosing environments that support, not sabotage, your clarity.

The Practice: Build Your Neural Trigger Map
Ask yourself:
- What sights, sounds, people, and tasks **rev you up or shut you down**?
- What moments consistently spark **anxiety, calm, confidence, or frustration**?
- What **times of day or types of tasks** feel naturally aligned - or completely draining?

Write it down. Use colour codes, sketches, or notes on your phone. You're not trying to fix anything - you're trying to **see the shape of your own brain's rhythm**.

Future-Proof Your Brain Insight

You can't regulate what you don't recognize. And you can't design a life that supports your brain until you know what overstimulates it.

"When I know what sparks or calms me, I stop reacting. I start responding."

To future-proof your brain, build a map of your **triggers and balancers** - and honor it. Because clarity about your wiring is the first step to **conscious reconfiguration.** This deflates the social narrative that we all will react positively to weighted blankets, blackout drapes, and white-noise machines. It makes room for diversity of thought and life dynamics.

Way 21: Empathy – The Super Power

Practice cognitive empathy: imagine another's perspective daily. This is a transformative way to future-proof your brain and your relationships. It's easy to say, "I'll try to put myself in your shoes," but most of the time we fail at this in an epic fashion – we're far too consumed with filling our own shoes!

Most of us have looked at someone who's experienced a tragedy, recurring disappointment, unfortunate circumstances repeatedly and marveled at the resilience of that person. How did they manage to go through such turmoil, chaos, and awful experiences and still function? Even more surprising and puzzling might be how they were able to thrive during this tumultuous time in their life. Where, you ask yourself, does that strength come from? An unkind voice in your mind might even say *"Why don't I have that ability??"*

Undoubtedly someone you know has experienced losing a job, ending a relationship, finding out they are ill, hearing about the death of a loved one…How do you feel when you hear their bad news? A perfectly normal reaction is to then reflect immediately on how fortunate you are not to be going through what your friend is experiencing. You might even think about a time when you felt similarly awful and hopeless - worse, you might choose that particular time of being confided in, to share how resilient you are.

"See," you say kindly, "I made it through an even worse ordeal than you did, and I'm perfectly fine!" As you say those words to a victim or a survivor of a traumatic experience - you inevitably pivot the focus onto yourself. The message you're broadcasting is **"I could handle it – so can you, so stop whining!" or "my pain is exactly like yours!"**

Alternately, you may find yourself justifying that if only you had the money, the support network, or the luck that they did - you too might have been able to manage what came your way. Human nature invariably results in comparing our misfortunes to others - the axiom *"it could be worse"* compels us to look outside of our experience and find who else has it worse than we do, or to recall a terrible time in our own lives where we did have it worse. *"I know exactly how you feel"* is how we may choose to express empathy. This creates a cycle of toxic comparison. We are inadvertently seeking to maximize or minimize our suffering in relation to the pain and suffering we witness others undergoing.

It is this very innate human pattern to anchor and compare that is rooted in our physiology. We hear traumatic news, we frame it and anchor it in our own experiences, leveraging our handy hippocampus for memory and amygdala for emotion. Hopefully, most of us possess an ingrained sense of compassion and ability to comfort others - but often despite our best efforts we struggle to respond to the troubles of others in a truly mindful or even genuinely comforting manner.

Our North American society celebrates lists and rankings - measuring and evaluating. When you go to the Emergency room - you're triaged, and you rate your pain on a scale. Pain and suffering are rated – just how "bad" does someone have it? These very thoughts and ideas diminish the individual nature of how we cope with trauma and stress. For a single mother – 90-hour work weeks and packed schedules where we sleep an average of 3-4 hours a night? This may be normal – she might even feel that she functions best in that type of time ecosystem. I know for me this was a huge badge of honor – a battle scar, if you will.

At work, we share anecdotes of our boss's derailment behaviours, the politics we must navigate - human beings very much enjoy complaining, it has a certain inexplicable joy. Dennis Prager, the famous American talk show host famously said the following words: *"Complaining not only ruins everybody else's day, it ruins the complainer's day, too."*

Complaining and observing others failing might even be ranked as two of our most enjoyable activities. The popularity of reality TV can possibly be ascribed to this unpleasant tendency of ours - we are fascinated by the tragedies of those around us. The German word **schaudenfreude** means taking pleasure in the misfortune of others. This doesn't mean the world is full of sadists, but researchers at Princeton (Cikara, 2013) have observed through a series of experiments that this tendency to enjoy watching others fail is quite normal.

Indeed, framing and anchoring are both approaches to presenting information to subjects in many psychological experiments. When subjects are given information and how they conceptualize it is determined by many variables: their mood at the time, the environment they are in, what are the alternatives presented to them, whether there are negative consequences to themselves or to others, and the order in which data is shared with them.

🧠 Future-Proof Your Brain Insight

Empathy isn't just about kindness - it's about **cognitive expansion**. The more you can hold another person's pain without inserting your own, the more room your brain makes for nuance, complexity, and connection.

"When I sit inside someone else's perspective - without comparison or correction - I teach my brain how to expand instead of defend." To future-proof your brain, practice **cognitive empathy** like you would a language: Daily. Deliberately. Without needing to be fluent - only willing. Try this the next time someone shares a disaster: ask "what do you need from me most right now? Distraction? Humour? Comfort? Help me help you in the way you need it the most" – constructive empathy is all about delivering what the other person needs, not what you need when you are suffering.

> *"Empathy is the only human superpower-it can shrink distance, cut through social and power hierarchies, and transcend differences."*
> *– Elizabeth Thomas*

Theme 2: Critical Thinking (and no I don't mean using Generative or Agentic AI)

Way 22: The Copycat

Change changes us, of course. But what happens when we change and moderate the way that we think? A mirror neuron is a neuron that fires both when we act and when we observe the same action performed by another. They're formed in the very earliest months of our lives, usually before we turn a year old. It's why babies delight in mimicry, because it is the most rudimentary form of learning. When we separate into current-day tribes like the families and social groups we're used to, we reinforce those neuronal patterns in yet another way. This insight is crucial to the future of human physiology.

- We need to learn how to practice diversity in our relationships in order build neural resilience and prevent white matter shrinkage. As white matter shrinks your brain becomes less effective in its communication and wiring – sticking to your own tribe is the best way to shrink your mind and world.

- Seeking out people who have different lived experiences, in terms of culture, gender, religion, belief systems can challenge our mirror neurons.

- Actively creating an environment for our brains where they are constantly exposed to different tribes is a choice that protects our brains' health. This practice contradicts every evolutionary mechanism possible – the eons long practice of seeking out the familiar in the name of safety and security.

🧠 Future-Proof Your Brain Insight

Familiarity may feel safe - but it keeps your brain rehearsing the same moves, repeatedly. True resilience comes not from agreement, but from navigating differences with curiosity, not fear.

"Every time I step outside my comfort zone of sameness, I give my brain the chance to grow beyond survival - and into connection."

To future-proof your brain, seek out minds, stories, and tribes unlike your own - not to agree, but to understand. Because **difference is not danger**. It's the deepest evidence that your mind is still capable of expansion. Mirroring and mimicry will always have a role in human learning and development but if only mirror neurons are firing, we become lifelong "copycats" – still functioning at the level of a toddler. This mimicry also has begun to show up in Generative AI-Human interactions, where many users begin to subconsciously start sounding like the algorithms they interact with all day. Intellectual levelling is what happens when large-language models group users into persona cohorts and those same users become trained to use a certain limited lexicon constrained by A.I. language sets.

> *"We become not a melting pot but a beautiful mosaic. Different people, different beliefs. Different yearnings, different hopes, different dreams."*
> *- Jimmy Carter*

Way 23: Navigating Health Crises

There are always going to be times when it becomes harder to invest any energy in building a brain that's ready for the future. This is especially true when we're facing a health crisis. Our bodies may feel or look different because of illness, and our brains are going to be constantly adapting to new versions of "normal". At 23 years old my ENT surgeon attempted to perform a tympanoplasty on me, it was his second attempt after first trying when I was 19 and finding an unusual growth while he was trying to repair my eardrum.

In that fateful second surgery he didn't expect to discover another growth attached to my facial nerve. I had exhibited none of the symptoms of what is called cholesteatoma. In removing it, the surgeon accidentally punctured my facial nerve, and I woke up at aged 23 with half of my face paralyzed – and no doctor who could tell me if I'd ever recover full muscle control.

Everyone around me said "sue him," instead I followed my father's advice and invested all my energy into healing and facial physiotherapy. Eventually I regained most of my muscle control – but even today the muscle impact is noticeable if you know how to analyze my facial expressions.

In these scary health moments, there are new pathways forward, especially because our brains are already challenging themselves to keep up with new information – new ways of moving our bodies and adjusting how we navigate environments as "differently abled" human beings

- After facing health or illness, it's critical to examine how our identity has changed and what our new needs may be, particularly when we experience physical changes.
- Spending time engaging in low-urgency tasks, like baking or painting, without any focus on perfection or accuracy, but with a focus on enjoyment, can help to soothe rapid neural changes
- We may find comfort stimulating our brains with new experiences, learning something new that uncovers a new part of our post-illness identity that is positive
- Finally, body identity reunification is a practice that involves bombarding your vision center with images of your "new" self. That same feeling of discomfort you get when you hear yourself speak, or see yourself on video? That's a result of the dissonance between how you imagine

yourself and reality. The major body dysmorphia that people often feel after illness means reprogramming your internal image.

Proprioception is our awareness of our physical self in relation to the world around us. It's how you know how high to raise your foot to navigate stairs. After a health crisis imagine yourself as a teenager – whose limbs are growing faster than your brain can keep track of. Invest in rebuilding your proprioception map to align to your new sense of physicality.

Future-Proof Your Brain Insight

Your brain is constantly trying to reconcile who you were with who you are now. But survival doesn't mean going back. It means learning to recognize yourself again - even when everything feels unfamiliar. *"This is my body now. This is my brain now. And both are still teaching me."*

To future-proof your brain after illness or major physiological change, practice **gentle self-recognition**: not forcing a return to the old self - but giving the new self a chance to feel *real*. Because reuniting with yourself after transformation is not a regression. It's a reintroduction. And your brain is wired to meet you there.

> *"It's not the strength of the body that counts, but the strength of the spirit."*
> - *J.R.R. Tolkien*

Way 24: Re-architecting Self-Perception Through Job Loss

How we perceive ourselves and how we believe that we are perceived by others influences our self-worth. Building a strong foundation of self-worth relates to two parts of the brain: the hippocampus (our memory center) and the medial prefrontal cortex (where we organize our thoughts and make decisions, including social decisions). Ballistic interruption activities can increase the capacity of our hippocampus, interrupting our tendency to either fight with others or run away from friendships when we think others don't like us. It all comes back to reinforcing the connection between that "old brain" and "new brain" – especially during times when we are feeling threatened and insecure – like after being laid off from a job.

Over decades of corporate life, I've experienced layoffs and also been the one signing off on restructuring. During a layoff your deepest sense of self-worth is attacked – the way you define your value in a North American context is so intrinsically tied to what you do for work.

- It's important to avoid becoming a recluse; we can maintain social interactions by changing the context and responses we have to social stress
- Connecting with others who have experienced reputational attacks and journalling about our feelings can help us shift our typical neural patterns of shame and embarrassment
- Exploring physical activities that allow catharsis of our anger such as boxing and axe-throwing can increase healthy cortisol and adrenaline cycles
- A new practice for managing stress post lay-off or termination is called "persona mapping." It begins with conducting an empathy interview with yourself – who are you? Who are you in relation to your work? Exactly how much of your self-worth is defined by your career? As you build this "persona map" and visualize it – the brain begins to segment out your self-worth from your career. Especially in current times, in the uncertain job market where job security is lower than ever before – persona mapping is a crucial tool for creating neural circuitry that is future-proofed.

🧠 Future-Proof Your Brain Insight

Your brain links self-worth to social safety - and when work is lost, your identity often goes with it. But losing a role doesn't mean losing yourself. It means learning to **locate your value beyond your title**. Abraham Maslow's hierarchy of needs explains beautifully why when our basic needs like food and shelter aren't met, it's almost impossible to think of more nebulous concepts- like the meaning of life."

"My self-worth isn't gone - it's not tied to how productive or successful I can be." To future-proof your brain, use tools like persona mapping to separate who you are from what you do. This isn't about letting go of ambition - it's about building a more **flexible identity**, one your brain can carry forward through *any* change, not just the ones you choose. Because your brain isn't wired to grieve a job - it grieves the loss of place, purpose, and permission. Give it new anchors, new maps, and let your worth expand - not shrink - in uncertainty. Play with the idea of devising a new answer to that ubiquitous social question: "So, what do you do?" I try this prompt during networking: "Tell me about who you are without mentioning work."

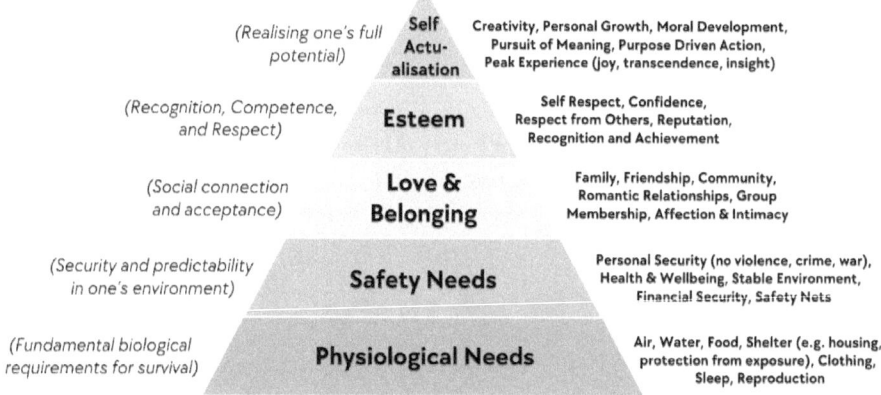

Image: Abraham Maslow's Hierarchy of Needs

Way 25: Make Peace Your Preferred State of Being

In the journey we take to future-proof our brains, happiness is an essential part of the equation. Interoception is the art of bringing awareness to our internal physical self to influence our emotional state. If proprioception is your awareness of self in relation to your external environment – think of interoception as your ability to monitor, track, and influence internal mechanisms (such as hunger or thirst). It's about using all our senses to examine how we're feeling, and not only emotionally, but with a focus on physicality – such as how fast our heart is beating at any given moment. There is a neural source of joy connected to the simple tasks of life. Creating a foundation of joy and long-term happiness begins with prioritizing neural mechanisms that result in serotonin. Serotonin has historically been less sexy than dopamine – and this makes sense, because dopamine is all about short term rewards.

- From functional activities like increasing cardio to seeking out more sunshine, to making dietary changes like more fiber and tryptophan, pumping up serotonin is a significant part of creating long term sustainable joy.
- Rerouting our neural synapses away from their focus on physical safety towards psychological happiness is possible
- Building mental associations between our reward systems and a sense of calm can result in increased joy, satisfaction, and authentic happiness.
- "Habit programming" is part of why new year's resolutions have an 80% failure rate – it all harkens back to understanding that the success of goals is often about evolution. Historically, winter was a time for conservation, while spring was for growth and fall for harvest – a new strategy of aligning tasks to seasons is helping people to increase their goal success by a minimum of 65%.

🧠 Future-Proof Your Brain Insight

Dopamine screams. Serotonin whispers. But it's the whisper that keeps you steady. This isn't biohacking – it's reconditioning your brain to accept peace as status quo.

"My joy doesn't have to be loud to be real."

To future-proof your brain, build habits that feed your calm, not your cravings. This means anchoring your life in the rhythms of movement, sunlight, sleep, food, and self-awareness - not to chase a high, but to build a foundation. Happiness that lasts isn't an accident - it's an **internal ritual**. And it starts by listening to your body when it says, *"this feels good, let's stay here a while."*

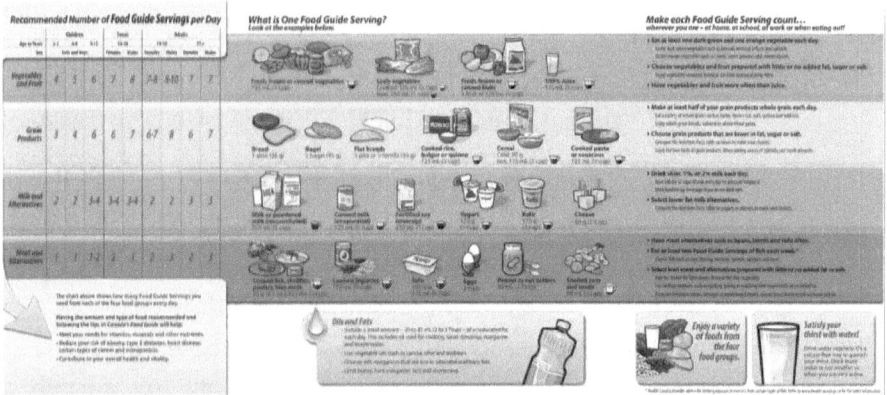

Image: Canada's New Food Guide, 2019 – an excellent place to start for building habits that feed and fuel your mind.

Way 26: Hack Your Human Blueprint

There is a massive shift occurring among all of us. Our gene diversity, and genetic expression, namely how our bodies use the genes we have, is growing in a significant way. Genetics are changing our brains, and how we adapt over eons, not just lifetimes. The idea of a perfect set of genes no longer exists, as life expectancy increases, along with treatments for diseases and a more effective understanding of our bodies and minds. Natural selection remains an important theory in science, but how it functions in current times has vastly transformed – trait selection and how we value traits is no longer according to "the strongest survive."

100s of years ago children with multiple sclerosis or cerebral palsy, or even premature babies had a very high infant mortality rate. Today – the genes that are passed down by each generation aren't always determined by which children are the most physically strong. As medical science speeds forward, the genetic cocktail of humanity continues to shift. The most adaptable are the ones to thrive. Cognitively adjusting is the new brute strength – the new prized immune system response.

- Because our neural footprints have evolved to preserve different genetic traits over time, we are the source of a range of potential traits.
- Epigenetics, namely the factors beyond the genetic code that inform how we turn out as human beings, can also influence change.
- We can confront our brains with the facts of genetic traits as adaptive and equally valuable.
- The practice of Epigenetic Exploration includes awareness of both genetic predisposition and the ability to influence gene expression through neural resilience training. How does it work? Let's assume your family has a history of immigration and settling in cities – awareness of your COMT gene impacts your control over it. The COMT gene ensures a dopaminergic response to urban living – it means that invariably people with this gene will be happier in cities.
- Mapping out our genetic memory and the associated behavioural components functions as a sort of neurological roadmap to understanding another dimension of why we feel more aligned to certain environments.

💬 Future-Proof Your Brain Insight

You are the first generation with the tools to observe, influence, and redirect your own biology - in real time. Access to life science knowledge, MRIs, EEGs, and unlimited digital libraries of open-source research.

"I'm not just shaped by evolution - I am continuing it."

To future-proof your brain, treat your traits as data, not identity. Understand your wiring, then design the life - and environment where your biology can *thrive*. Because the future of evolution isn't about strength. It's about neurological self-awareness. Human beings need never be completely at the mercy of environment or genetics – both play fundamental roles in our evolution, but it's also possible to influence our physiology towards resilience and grit to decode your personal genetic trajectory and architect a roadmap of where to start rewiring towards.

Understanding the nuances of your neurological Lego blocks means creating a project plan for cognitive renovation. It means identifying your natural or acquired skills with the highest probability of making you the most successful in a particular environment.

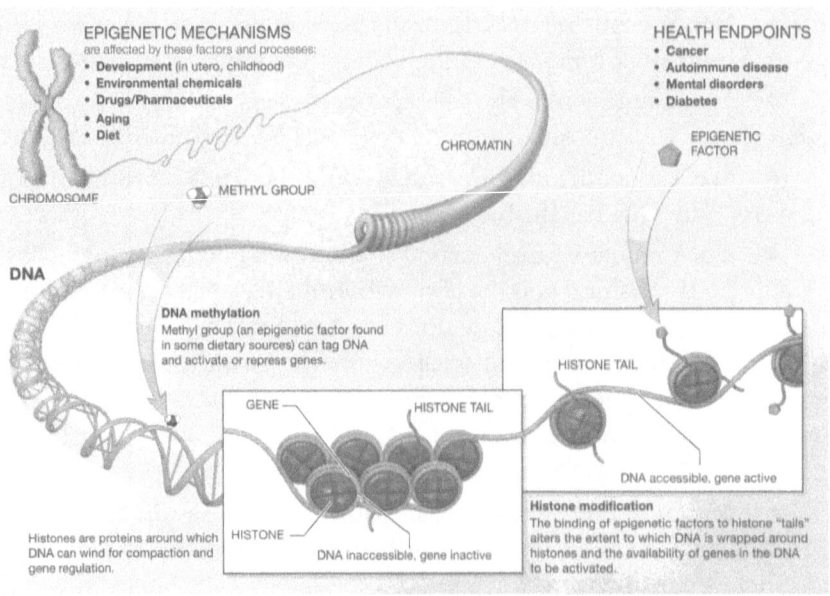

Image: National Institute of Health – Epigenetic Mechanisms

Way 27: Build Your Bionic Brain

What does the next iteration of the human brain look like? Extensive exposure to technology has permanently changed multiple regions of the human brain. Changes in our environments matter as well. Neurotechnology and neural implants may shift our medical responses to brain trauma. From eliminating neurotoxins to providing information on neural function back to an individual via wearables, or nanorobotics alleviating brain inflammation. We can even enable rehabilitation of traumatic brain and spinal cord injuries.

- Imagine a world where all our neural data and ability was on display for us, and we could make more informed choices about our neurological futures
- As the human brain continues to evolve, technology plays a crucial role in this evolution and means that the majority of this growth will be in the occipital and frontal lobes (vision and critical thinking)
- Knowing more about how our brains work, we have the power to shape our own futures. Continuing to balance the utilization of GenAI, agentic AI and metacognitive intelligence is imperative as we strive to maintain neural evolution and growth in the "new brain" – preserving the invaluable skills of creativity, play, and critical thinking. More than ever before, our ability to defy patterns, and challenge norms is where the rarity of human thought and behaviour defies understanding, predictability, and machine learning models. The brain goes through a massive change at every Industrial Revolution – the widespread use of cell phone is just one example where technology has impacted our motor cortex, auditory cortex, and visual cortex.

Future-Proof Your Brain Insight

The next version of your brain won't arrive in a lab - it will emerge from the choices you make right now. How you think. What you resist. When you play. Whom you trust. These are acts of architecture.

"I will evolve - but I will not erase the parts of me that make me human."

To future-proof your brain leverage technology not just to conform and upskill, but to amplify what makes you unpredictable: curiosity, creativity, contradiction, and conscience. Because *those* are the signals no machine can replicate - and they are your most valuable data. Are you using technology to upgrade your functional cognitive capacity or to get more done without always doing more? One of the glaring gaps in AI literacy courses is that people are not taught how to train the AI tools they are using. Learning and development courses focus on task completion rather than on explaining how AI models function and evolve. A brain that can balance using tech with controls, avoiding screen addiction and adverse neurochemical responses is one that takes ownership over the impact of technology.

> *"I think the biggest innovations of the 21st century will be at the intersection of biology and technology. A new era is beginning"-Steve Jobs*

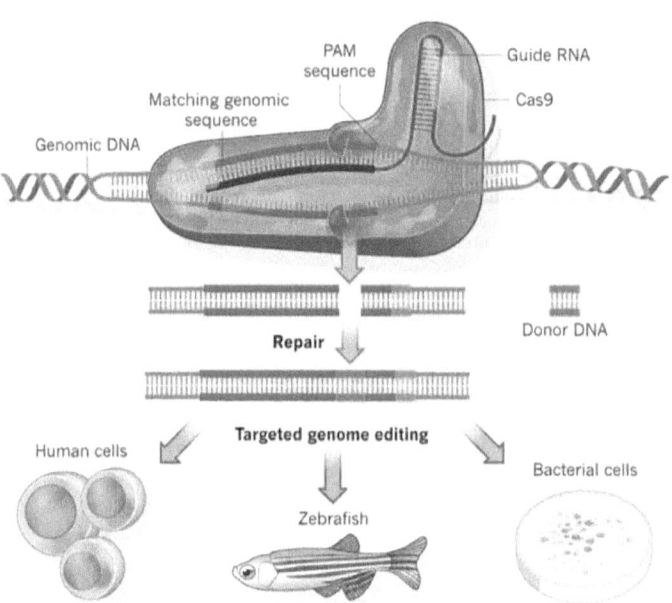

Image: CRISPR Cas9 Genome Editing System – utilized to precisely edit RNA and modify specific genome sequences.

Way 28: Match Your Brain to Your Environment

You might remember earlier in this book, we touched on a gene called COMT - the one that governs how your brain handles dopamine, and in turn, how you respond to stress, stimulation, and novelty. Let's come back to that now - because this tiny stretch of genetic code may hold one of the most powerful keys to understanding why you feel at home in some places and lost in others.

COMT comes in a few variations, but the most well-studied are Val/Val and Met/Met. These variations determine how fast your brain clears dopamine - a chemical that regulates everything from your focus to your sense of pleasure to your emotional bandwidth.

- If you're a Val/Val, you clear dopamine quickly. That means you're likely energized by fast-paced, high-stimulation environments. Cities likely make you feel *alive*.
- If you're Met/Met, your dopamine lingers longer. That means you may feel overstimulated by chaos and noise - and you crave calm, structure, and familiarity.

What's extraordinary is how these genetic traits - formed over millennia - **still shape our modern mental health**. Your nervous system isn't just reacting to your to-do list. It's reacting to the *architecture* around you. The light. The sound. The people. The rhythm. And that brings us to a profound truth: **Environment matters. Deeply.**

The Urban Myth of One-Size-Fits-All

We've been sold a myth - that there's one kind of dream life, one kind of "success," one kind of ideal neighborhood. For some, it's the buzz of downtown. For others, it's the quiet of tree-lined streets. But when your environment doesn't match your brain's operating system, you feel it - in the form of:

- Chronic irritability
- Emotional burnout
- A flat or anxious mood
- Constant low-level stress

And it's not just "stress." It's *neurological friction*.

Your brain might be screaming for space - or for more stimulation. But you've been taught to ignore those signals. To push through. To "adapt." To stay where it looks good on paper. We can do better. This is

your invitation to listen more closely to your lived experience. You don't need a genetic test to tune in. Ask yourself:
- Does the pace of my environment match the pace of my mind?
- Do I leave my space recharged - or depleted?
- Am I holding tension I don't even realize I'm carrying?

Because the future of mental health isn't just internal. It's *architectural*.

🧠 Future-Proof Your Brain Insight

The best version of your brain doesn't emerge in a vacuum. It emerges in the **right habitat** - one that matches your wiring.

"My mental health is not separate from my environment. It's shaped by it."

To future-proof your brain, stop blaming yourself for what your body is trying to signal. Pay attention to the environments where you feel most *you*. That's not indulgent - that's biology. And it might just be the beginning of coming home to your own nervous system. Eschew the narrative that every person thrives within the confines of a "9-5" workday, going home to a spouse and children, comfortably ensconced in an abode surrounded by a white picket fence. Even the supposition that we all need the same amount of sleep has been belied by neuroscience. Cognitive alignment is less about matching to one predefined human roadmap of perfection, and more about figuring out why your particular brain was designed the way it was.

Theme 3: AI, we know it...we have a Hate/Love Relationship with IT (pun intended)

Way 29: Do it Yourself

During my elementary school years in Canada, we were forbidden from citing websites in our book reports. Teachers insisted that "real" research came from printed sources - books, journals, encyclopedias. It felt like a punishment at the time, but I can still remember the feeling of discovery, walking through aisles of hardcovers and microfiche drawers, stumbling on tangents and contradictions I never would've searched for on purpose. That process - the slow, analog sifting of data - taught me something that AI will never replicate:

How to think critically. How to connect the dots. How to sit in not-knowing. Today, we're speeding past that. We prompt. We scan. We accept. And the more we let AI answer questions for us - without interrogating or diverging from the path - the fewer opportunities we have to build the deep cognitive networks that fuel long-term intelligence. 80% of the books that I'd comb through in my research efforts would lead me down a rabbit hole that didn't even correlate to my hypotheses. Yet the vast spider web of knowledge in my mind would inevitably grow: my fact-checking and scientific method skills flourished, preparing me for a future of PhD Research and life as an AI Neuroscientist (a career that certainly didn't exist when I was first asked what I wanted to be when I grew up).

🧠 **Insight: Independent research activates critical thinking circuits in the prefrontal cortex and strengthens long-term memory.**

When you seek answers yourself - whether in a library, through fieldwork, or even by comparing conflicting sources - your brain engages:

- **The dorsolateral prefrontal cortex** (analytical reasoning)
- **The anterior cingulate cortex** (conflict resolution)
- **The hippocampus** (long-term memory integration)

In short, doing your own thinking builds stronger, more complex, more resilient neural networks. Letting AI "fill in the blanks" might feel efficient - but it's neurologically lazy.

🧠 **Insight: The "hunt for data" strengthens adaptability, contradiction tolerance, and intellectual flexibility.**

When you find a piece of information that contradicts your hypothesis, your brain has no choice but to recalibrate. That recalibration is called cognitive flexibility - and it's one of the most critical skills in a world shaped by uncertainty and exponential change. If you're using GenAI to only present you with data that reinforces your perspectives and presuppositions, you're killing brain cells.

Outsourcing thinking removes the most valuable part of the process: *the rewiring that happens when your own mind finds, filters, and refines information.* Of course, AI has its place. It can brainstorm, organize, suggest. But it should never become your default mode of cognition. The moment you stop engaging with the process, you stop building the architecture.

🧠 **Future-Proof Your Brain Insight**

Answers don't strengthen the brain - effort does. AI can supplement your mind, but it should never **substitute** your mind. *"Every time I choose to think through a problem myself, I build a brain that can handle more than answers - it can handle ambiguity."*

To future-proof your brain, stop chasing fast answers. Slow down. Dig deeper. Seek data that challenges, not just confirms. Because your most powerful neural upgrades won't come from clicking a prompt - they'll come from the struggle to make meaning on your own.

In 2024 my PhD Research examined a total of 1,923 adults across Canada (n = 946) and the United States (n = 977), ages 25–57 (M = 38.4), who completed ten GenAI-assisted tasks designed to simulate common executive demands, including planning under uncertainty, multi-step sequencing, decision-making, and reflective reasoning. Participants were encouraged to use commercially available large language model systems as they normally would. Measures included prompt activity, time-to-decision, override behaviour, and Likert-scale assessments of AI reliance, confidence in reasoning, and perceived cognitive autonomy.

Across tasks, 58% ± 7% of participants agreed that "AI did most of the thinking," with higher reported offloading during planning and sequencing tasks. Greater prompt dependence and lower override frequency were associated with reduced self-reported confidence in independent reasoning (r = −.61, p < .01). Qualitative responses highlighted recurring themes of cognitive outsourcing, diminished perceived ownership of ideas, and trade-offs between speed and depth of thought (Baldeo, in press).

This research yields hopeful insights for proof of increased activation in what we refer to in neuroscience as "Brodmann areas," which are distinct regions of the brain broken down into 52 segments that correspond to different cognitive functions. When AI was used correctly, Brodmann areas 6, 10, 32, and 46 spiked in activity – so what does this mean to the non-neuroscientist? It's powerful data, because these regions possess the highest concentrations of "pyramidal neurons," which are the most evolved types of brain cells possible. So, it's worth pausing on the pervasive conclusion that ALL AI use is wretched for human beings!

Way 30: The AI Study Buddy/Therapist?

In 1966, MIT computer scientist Joseph Weizenbaum built one of the first AI systems: **ELIZA**. In 2023, I was fortunate to visit the Computerspiel Museum (a wonderful collection of every video game and stage of computers) in Berlin and interact with ELIZA in person. It's a giant computer with the typical black screen and green type that you see in "old" movies. It was a simple program designed to simulate a therapist by repeating back user responses in slightly reframed questions. It was primitive. It was basic. And it worked - disturbingly well.

Weizenbaum's goal wasn't to replace therapists. He wanted to demonstrate the illusion of empathy, and in doing so, highlight how easily humans project contrived emotional depth onto machines. To his dismay, users began forming emotional attachments to ELIZA. Some even confided in it - believing they were being "understood." Even in 1966 AI delusion and hallucinations were happening. Weizenbaum was horrified. He saw this not as progress, but as proof of how quickly we would turn to machines to fill human needs. Again, keep in mind this was 1966 – **AI IS NOT NEW!** Fast forward to now, and we're using GenAI to write essays, process breakups, make stock picks, brainstorm career moves, and even emulate therapy. But the question remains:

Where's the line between assistance and substitution? Emotion AI is the category of AI that focuses on interpreting, analyzing and in some cases even responding to human emotion. It's been used in certain parts of the world in immigration interrogation to determine if detained individuals are being truthful. The problem – people's facial expressions, tones, vocal cadence and ways of expressing themselves are vastly different across cultures and age groups.

In 2015 I worked with a company that was excited to start using facial recognition in virtual interviews. Their selling point was that they would track the candidates' facial expressions and be able to determine honesty using behavioural analysis such as pupil dilation, making eye contact, and eye movements. The platform eventually failed, because these behavioural traits are extremely individualized and certainly never account for nuances such as neurodivergence for example.

🧠 **Insight: AI can support thinking - but not substitute emotional processing or deep learning.**

Using AI as a **study buddy** can be brilliant:
- It can quiz you, summarize texts, or explain concepts in new ways
- It can help you test your understanding, explore opposing views, or rehearse arguments.

But here's the danger: If AI becomes your default cognitive partner, you lose the effort - and the emotion - that makes growth meaningful. Sure, you could give a tool a prompt to "argue" with you – but human unpredictability is hard to well….predict.

Learning and healing both require:
- **Struggle** (which builds neural plasticity)
- **Reflection** (which integrates memory and emotion)
- **Context** (which AI lacks entirely)

Therapy isn't just "talking." It's co-regulation. Empathy. Silent nuance. A nervous system recalibrating with another. No chatbot can hold that space. This doesn't mean don't use the tools. It means use them *with boundaries*. Let AI be a study buddy - not your healer, your moral compass, or your emotional crutch.

Because what AI mirrors, it doesn't metabolize. You still need to do the thinking. You still need to feel the feelings. And keep in mind that YOU are training your GenAI interface – it possesses YOUR inherent bias. Consulting AI as a tool during therapy can be revelatory, but firing your human therapist as a result? Not advisable.

🧠 **Future-Proof Your Brain Insight**

AI can reflect, but it cannot relate. It can simulate, but it cannot hold space. *"I'll use the tool. But I won't let it teach me to bypass the work."*

To future-proof your brain, draw a boundary around where AI belongs - and where only your effort, emotion, and lived experience should live. Because some parts of learning - and healing - must be human.

> *"Psychotherapy isn't a twentieth-century artifice imposed on nature, but the reinstatement of a natural healing process."*
> - *Patricia Love, Author of "Emotional Incest"*

Way 31: Journalling, By Hand

For years, I resisted journaling. It felt self-indulgent. Too intimate. Too "Dear Diary." Even when my partner suggested I start journalling because I barely made time for therapy, I resisted for years. I told myself: "I'm not a teenager with a diary! I'm a grown woman!" I worried that writing down fleeting emotions might solidify them - give them more weight than they deserved. Then I tried it my way.

I bought a structured workbook: *Let That Sh$t Go** by Monica Sweeney - a mix of prompts and irreverent humor that made self-reflection feel manageable. I picked a secondary free-writing journal with a Monet print on the cover - beauty matters when you're opening your mind. And then there was my list-making notebook: a chaotic place where everything from anxieties to grocery items spilled out, side by side.

Journaling didn't fix everything. But it *did* change something: The act of slowing down. The pause. The hand-on-paper awareness that something inside needed translating. Yes, I still talk to AI. Sometimes I use Claude for emotional processing when my therapist isn't available. I use ChatGPT for science and structured research. But no matter how helpful these tools are, nothing replaces what happens when your own hand tries to catch your own thoughts. Because journaling isn't just venting. It's organizing. It's sense-making. It's a physical act of presence.

🌀 **Insight: Handwritten journaling activates different neural pathways than digital input.**

It strengthens memory, emotional regulation, and self-awareness. And most importantly - it's yours.

- Writing by hand builds cognitive resilience through motor function and introspection
- It creates space between thought and expression - a buffer zone where insight can emerge
- It doesn't require optimization, feedback, or output - it just needs honesty

AI is helpful. It's even companion-like. But until it becomes sentient and reached metacognition, it will never know how it *feels* to sit in silence with yourself. To stumble through your own messiness. To write something raw and leave it there, unedited, scribbled and imperfect.

Journaling is the opposite of optimization. And that's exactly why it works (and exactly why I found it so difficult initially).

🧠 Future-Proof Your Brain Insight

Let AI be your co-pilot, not your confessor. Use it to expand your thinking - not escape your feeling. Your journal isn't just a page. It's a mirror. And some reflections should be drawn in ink, not pixels. *"I'll use AI to process. But first, I'll check in with the only voice that really knows what's going on - mine."* Journalling is an overdue reminder that you know exactly who you are, even when it feels like that is the farthest reality from the truth.

> *"Journalling is medicine, it is an appropriate antidote to injury."*
> - *Julia Cameron, Author of "The Artist's Way"*

Way 32: Be Multi-Source

It's tempting, isn't it? To let your feed become your fact-checker. To scroll TikTok or Instagram and assume the algorithm's got your back – to assume that what's popular must also be true. But popularity is not the same as credibility. And the easiest stories are often the least complete. Our brains were designed to triangulate - to observe, compare, weigh, and discern. Not to consume and conform. 60-second click-bait news on social media feeds destroy more than common sense. A 2024 study of short-form videos revealed that short-video addiction has a risk of diminishing self-control and executive function (Yan, 2024).

In a world where AI tools are everywhere and headlines are optimized for clicks, your job is to stay curious - and cautious with every new tool you are exploring, without buying into fearmongering. I use ChatGPT for context. Claude AI for nuance. Perplexity AI for summaries. And when I'm lucky, Manus AI for a rabbit hole or two. Each one has its strengths. Each one has its blind spots.

In August 2024 a journalist at a Wyoming newspaper called The Cody Enterprise admitted to using AI to generate articles and stories – but he was only caught when a competing newspaper realized bizarre inaccuracies in his journalism. Without fact-checking the stories would've continued, and in this case the newspaper thought they had the most modern AI-tracking tools in place.

🧠 **Insight: Truth isn't found in one feed, one source, or one voice. It's constructed through contrast.**

To future-proof your brain, you need to be a *critical curator*, not a passive consumer.
- Diversify your news diet: old-school journalism, long-form podcasts, subreddits, AI assistants, international outlets
- Ask different tools the same question - notice what's repeated and what's omitted
- Seek disagreement, not just reinforcement

Because your brain doesn't grow from being right. It grows from being *stretched*. From the friction between ideas. From the tension of not knowing - yet. The most dangerous thing isn't misinformation. It's overconfidence in a single version of the truth.

🧠 Future-Proof Your Brain Insight

Start your day with reading a peer-reviewed journal, even if you aren't a scientist or researcher. Schedule time for podcasts, news headlines, and yes, even nonsensical TikTok videos. Train your brain to absorb insights and inspiration from multiple sources because over-reliance on one form of media in the Age of AI is putting you squarely at risk of Information Asymmetry with your peers. In an age of information abundance, cognitive laziness is the real threat. Switch tools. Cross-check. Stay uncomfortable. The more angles you gather, the clearer the picture becomes. *"I'll use the feed - but I won't be fed."*

"My biggest fear is that people will attribute fake quotes to me and millions of morons on the internet will believe it." - A quote attributed to Albert Einstein that he obviously never said…given that he didn't live to see the internet.

Way 33: Go Cold Turkey with Tech Sometimes

Remember those good old-fashioned whiteboard brainstorms? Pens squeaking. Ideas flying. Tangents welcome. There was no tab-switching, no pop-up reminders, no background pings. Just brains. And walls. And markers. And metaphorical headbanging against a wall. And forgetting to check your phone because you were busy solving problems with your brain!

If you were born before the internet took over - hi, fellow millennial from the 1900s - you might remember what deep focus felt like. If not, here's the good news: it's still accessible. But it is going to take a little courage. And a full system shutdown.

Cold turkey is exactly what it sounds like. No Google Docs. No Miro. No ChatGPT humming in the background, ready to autocomplete your next thought. Just you, your mind, and a physical space where creativity doesn't come with single sign-on feature.

🧠 Insight: Tech extends your brain - but it can also eclipse it.

Creativity isn't just what you *say*. It's what you *notice* before you say it. And sometimes, tech moves too fast for your deeper thoughts to catch up.

- Going analog slows you down - and that's a good thing.
- Physical brainstorming (whiteboards, sticky notes, paper) activates spatial reasoning and memory differently than screens.
- No digital tool can replicate the embodied, collaborative, often chaotic beauty of real-time ideation.

Tech will always be there when you come back. But give yourself the gift of *not relying* on it. Disengage so your original ideas can surface. Because sometimes, the best ideas aren't generated. They're *remembered*. From the quiet. From your gut. And this is coming from someone who has built tech for 20+ years.

🧠 Future-Proof Your Brain Insight

Digital tools are powerful. But so is your imagination - when it's not being distracted every seven seconds. Take one day a week to unplug your thinking. Uncloud your creativity. Unblock your brilliance. *"I'll use the tech. But I won't let it replace my mind."* As a developer and technologist myself, one item that is hugely neglected in IT standards is how the design of apps can harm users. We focus on features, functionality, throughput &

expediency. Setting screen limits for ourselves isn't juvenile, it's mindful utilization of tools. We would not endorse sitting watching television 24/7 for any human being – why should phone use be viewed any differently? "Couch potato" does not sound any worse than "phone potato."

"Concern for man himself and his fate must always form the chief interest of all technical endeavors. Never forget this in the midst of your diagrams and equations."
(Albert Einstein, speech at the California Institute of Technology, 1931) - <u>actual quote by Albert Einstein</u>

Theme 4: Stress, Resilience, and Gratitude

Way 34: Handwriting & Cursive – Bring it BACK!

How many people still know how to write in cursive? If you grew up with those thick handwriting workbooks in the '90s - the kind that alternated between big, dotted letters and winding cursive loops - you are part of the last generation that saw cursive as a skill, not a curiosity. Since the 1970s, cursive instruction has declined dramatically, especially in the U.S., where Common Core standards removed handwriting from the curriculum. The reasons are predictable: the rise of screens, keyboards, voice-to-text, and let's be honest - a bit of plain human laziness.

But here's what we've lost in the process: a form of cognitive engagement that activates multiple parts of the brain at once. When I joined *CTV's The Social* in October 2024 to talk about gratitude and neuroscience, I emphasized something we often forget: Writing by hand changes the brain and is the kind of gratitude that sticks. Expressing thanks verbally is good. But expressing it *by hand* - especially in cursive - does something different:

- It activates your **hippocampus**, the memory center of the brain, by embedding the emotion more deeply through fine-tuned physical motion.
- It lights up your **motor cortex, visual cortex,** and **language centers**, making the act of writing more **multimodal** than typing.
- It increases **neural integration**, which helps solidify the emotional resonance of what you're writing.
- And it creates a more durable memory trace, meaning you're more likely to remember the moment - and the feeling - in the long run.

Cursive: The Neural Signature

Unlike typing - which uses repetitive, uniform finger movements - **handwriting (especially cursive)** requires:
- Variable speed
- Directional changes
- Pressure modulation
- Continuous visual-motor coordination

In other words: it's *neurosensory exercise*.

Research shows that cursive writing enhances:

- **Working memory**
- **Literacy**
- **Self-regulation**
- **Processing speed** - especially in developing brains, but also in adults seeking cognitive longevity.

And because cursive is unique to each person, writing in your own hand may serve as a kind of cognitive fingerprint, reinforcing identity and agency in subtle but meaningful ways.

Future-Proof Your Brain Insight

Typing is efficient. But it doesn't embed as deeply as handwriting. Handwriting is slower - and that's the point. *"The slower the motion, the deeper the memory."* To future-proof your brain, give it moments of tactile, analog engagement. Write a thank-you note. Journal by hand. Practice your signature, even if it's messy. Every loop and flick of the pen builds stronger mental maps. Because memory isn't just about recall - it's about feeling things enough to remember them. The days of aimlessly scribbling on a lined-paper created space in the brain for motor cortex activity segmented from work product and that IS a brain upgrade.

> *"Letters are something from you. It's a different kind of intention than writing an e-mail."*
> - *Keanu Reeves*

Way 35: Say Thank you – Out Loud!

It sounds almost too simple. Too analog. Too... human. But if you want a sharper, more resilient brain? Start talking. Out loud. Especially when it comes to gratitude. In a world of DMs, thumbs-up reactions, and silent Zoom meetings, many people can go full days without using their voice - Not for connection. Not for kindness. Not for anything beyond "You're on mute."

If you live alone or work remotely, you know this. Silence becomes a lifestyle. And soon, gratitude becomes a thought - not a practice. But saying "thank you" out loud does something different. It lights up your social brain. It signals connection. It produces oxytocin - that feel-good, trust-building hormone. It anchors gratitude as a real, felt experience. As a born and raised Canadian, I've often heard the teasing stereotype that all Canadians learn to say, "thank you" and "excuse me" as our first words. In fact, my first word was "Don't!"

Whether you're Canadian or not, learning to say thank you out loud has a plethora of cognitive bonuses. And then there is the unavoidable added benefit of kick-starting two major areas of your brain into overdrive when you go beyond thinking to speaking (and of course listening): Wernicke's area (language comprehension) and Broca's area (speech production). The caveat here is that ever since the advent of OpenAI and ChatGPT more and more humans find themselves "conversing" with AI tools and indeed preferring this interaction over talking to another person.

Large-language models are trained on significant amounts of human conversational data. And a singular element of human conversation is how we learn to signal being likeable. At the simplest level, amicability is a social signal to build tribal bonds. We tend to like people who sound like us, and furthermore, agree with us. It is therefore supremely unsurprising that GenAI would be developed with a conversational tonality that strives towards similar aims.

You may notice yourself starting to sound like your friends or spouse. Perhaps you marvel at your children starting to sound just like you. I recall the first time my son said "Mother, you are not elucidating your thoughts calmly!" In Way 22 I first introduced this concept of mimicry and mirror neurons – this is evident not only in the human-human relationship but also within the human-AI relationship. In 2025 scores of AI users began to complain that GenAI tools were not conveying accurate information, that the tools were giving them "AI hallucinations." Many algorithms during that year ramped up what I refer to as the "amicability quotient" in their models. The overarching goal was to sound "more human." But one of the most crucial elements of a healthy human relationship is the ability to challenge each other to grow and evolve.

The success of AI-human relationship must be predicated upon certain boundaries – where we do not further anthropomorphize GenAI, where we preserve the importance of human-human interaction, and where users take ownership of their own agency in learning rudimentary processing skills and critical thinking first, before outsourcing those imperatives to GenAI. Secondarily, and no less importantly, preservation of cognitive longevity must be prioritized through the design of Conversational AI tools and proper user training.

Noam Chomsky, in his examination of Richard Lewontin's theories around linguistic evolution, raised the core question around how language itself can shape our scientific conceptualization (Chomsky, 1997). The language we use to describe GenAI today shapes the development of human-AI translations and how humanity perceives AI moving forward into the 8th Industrial Revolution. Preserving and advancing linguistic complexity, semantics, syntax, and referenceability towards a new form of human-AI language requires awareness that this interaction itself is creating a new lexicon and format which bidirectionally shapes human-human communication. This doesn't mean you need to say "thank you" to your GenAI tools, but it does mean you need to remember that how you speak to your fellow human should not mimic how you speak to AI tools.

Insight: Expressing gratitude out loud strengthens your emotional circuits, deepens social bonds, and even improves immune function.

Your brain can't fully *encode* a feeling if you don't *externalize* it.
- Verbalizing gratitude builds emotional regulation and empathy.
- Spoken words activate motor and auditory processing, reinforcing memory and meaning.
- Saying it out loud turns an internal state into a shared one - and humans thrive in shared states.

Future-proofing your brain doesn't mean uploading it to the cloud. It means grounding it in the ancient technologies of connection. Gratitude. Presence. Voice. So next time you feel thankful, don't just note it. *Say* it.

To your partner. To your coworker. To your barista. To your dog. Let your brain - and someone else's - feel the real-time resonance of being acknowledged.

🧠 Future-Proof Your Brain Insight

In a world of silent scrolling and typed thanks, the spoken word is your neural superpower. It is a medium that preserved your lingual function, your physiologically evolved ability to articulate your thoughts beyond rudimentary sounds. Speech is a gift that we occasionally forget to use in a world of text messages and social media notifications. Make it a practice to speak to other human beings face-to-face at least once daily – call your friends when you see them suffering – nothing can replace the experience of auditory and verbal connection. *"I'll feel gratitude. But I'll also speak it - like it matters. Because it does."*

> ***"Gratitude turns what we have into enough, and more; it turns denial into acceptance, chaos to order, confusion to clarity."***
> ***- Melody Beattie***

Way 36: The 4x4 Method...No it's not Algebra

Breathwork, like AI, is not new. And like AI is not taught early enough. As children, we hear "calm down" or "take a breath" – but we never learn in standard curriculums about how breathing influences brain wave patterns, lowers blood pressure, and increase oxygen to the brain.

The 4x4 Method sounds simple enough. Four seconds in. Four seconds out. Repeat. But behind this rhythmic pattern is just one example of the most powerful tools your nervous system has: control over your breath. The U.S. Navy SEALs call it box breathing. In combat, chaos is guaranteed, but breath - that is the one variable they *can* command. And it is often the difference between panic and presence. Box breathing - also known as the 4x4 method - trains your brain to override stress responses, regulate emotion, and restore clarity. And the best part? You do not need a battlefield. You just need a moment. At your desk. In your car. Between Zooms, or MS Teams, or Google Meets. Before a tough conversation. After reading the news.

Most of us are quite familiar with the advice to "just breathe" or "go meditate" but the 4x4 method specifically has been proven to override the limbic fight or flight response. Its power lies in the intense adjustment that occurs in a person's heartrate and the amount of oxygen being rapidly directed towards the brain (Acala, 2020).

For those of us who have experienced giving birth, we are well familiar with the power of breathing to help us in moments of intense pain! This strategy is known as Lamaze breathing and the efficacy is all tied to refocusing attention away from pain. Unsurprisingly, breathing has the distinct potential to skew brain activity. Oxygenation is what keeps your brain working.

It might sound obvious that the brain needs oxygen, but as a proponent of biological awareness allow me to share the science behind it: oxygen is what enables your cells to convert glucose into adenosine triphosphate (ATP). ATP is the primary power source for your neurons to engage in signalling each other.

🧠 **Insight: Intentional breathing activates your parasympathetic nervous system, downregulates cortisol, and sharpens cognitive function.**

In just one minute, your body can switch from fight-or-flight to rest-and-respond.
- Breathe in for 4 seconds
- Hold for 4 seconds (optional)
- Breathe out for 4 seconds
- Hold for 4 seconds (optional)
- Repeat for at least 4 rounds

Do it anywhere. Do it often. Do it especially when your brain wants to *do something else* - scroll, snap, shut down. This is emotional regulation in its purest form. No app required. No subscription needed. Just you. Your lungs. And a willingness to pause (which is harder for some of us than others).

🧠 Future-Proof Your Brain Insight

The future will reward thinkers who can stay calm in uncertainty. In the midst of battles, metaphorical and physical, there is rarely space to withdraw and collect oneself in a relaxed fashion. The brain didn't evolve in a calm and controlled lab. Train for that today. Start with your breath. *"I'll use my inhale as an anchor. And my exhale as a reset."*

> *"Breath is the bridge which connects life to consciousness, which unites your body to your thoughts. Whenever your mind becomes scattered, use your breath as the means to take hold of your mind again."*
> - Thich Nhat Hanh, Author of The Miracle of Mindfulness

Way 37: Resilience Playlists

Sometimes resilience does not look like grit or grind. Sometimes it looks like hitting "play" on a movie you've already seen 14 times. Turns out, there's science behind that comfort. Rewatching your favorite movies - or replaying go-to playlists - isn't laziness or nostalgia. It's *regulation*. In a world that feels chaotic, familiar stories create predictability. You know what's coming. You know how it ends. And that sense of control, even if simulated, soothes your nervous system. Research shows that predictable narratives can reduce cortisol levels, ease anxiety, and restore emotional balance. It's not about escape. It's about *restoration*.

Think about how you may react to returning to familiar environments from positive moments in your life, or perhaps why visiting your grandparent's home always calms your nervous system. There is a common idiom that says, "familiarity breeds contempt," (often attributed to author Geoffrey Chaucer and his work *Tale of Melibee, 1386*) but in the case of your cognitive capacity for resilience, familiarity can be a healing touchstone.

🧠 **Insight: Revisiting familiar music or movies activates brain regions linked to reward, safety, and emotional regulation.** This isn't just entertainment - it's a cognitive self-soothing tool.

- Your brain loves patterns - and familiar media delivers them
- Rewatching a movie creates space to process emotion without new stress
- Music triggers memory and mood states that can interrupt rumination or panic

So yes, make your resilience playlist. The one with that one song that always hits. The one with that chorus that makes you breathe deeper. Watch that old comedy again. Queue up that rom com you know by heart. For me it's a three-way tie between *Age of Adaline, Something's Gotta Give, and It's Complicated!* Whatever your favourite movie is, let it wrap around you like a weighted blanket. In a high-stimulus world, we forget that comfort can be a powerful reset for your parasympathetic nervous system.

🧠 **Future-Proof Your Brain Insight**

Your brain does not always need new inputs. And that fact is entirely at odds with so many quintessential experiences of our modern day lives -

social media feeds, YouTube videos, AI conversations.... Sometimes strength comes from return - from rhythm, repetition, and stories that already know you. *"I'll embrace what soothes me. Because peace builds power."* This may sound contradictory in the face of chapter after chapter about exploring novelty. Intelligence and neurological evolution is as much about change as it is about learning control over your emotional regulation mechanisms.

"Music expresses that which cannot be said and on which it is impossible to be silent."
- Victor Hugo

Way 38: The Gratitude Anchor

You don't need a gratitude journal to do this next exercise (and no this isn't a repeat of Way 35). Or a ten-point list. Or a perfect morning routine. You just need *one moment*. One memory. One anchor. Something that pulls you out of your spiral - and back into safety. It could be the way your kid laughed with a missing tooth. The warmth of your grandmother's hands and the way she wore a ring on every single finger, everyday. The night you felt seen, or free, or finally at peace. The relief after signing divorce papers or the happiness on the day you finally married someone who made you feel safe. A sunrise. A song. A stupid inside joke. This is your *gratitude anchor* - a single memory you keep in your back pocket. It's portable. Accessible. Instant. And when things get loud, fast, or overwhelming? You reach for it. It's your safety blanket. Picture yourself walking around with a bag of memories.

Reading through the pages of this book you might get the impression that this neuroscientist eschews all the functions of the limbic system. Memory is controlled by your hippocampus, which maintains a constant residence in your limbic system. The crux of the discussion on limbic responses is not that all brain function from our "ancient" brain is to be vilified, but rather that we can upgrade its utilization to be more aligned to our contextual environments in current day life. A memory anchor that surrounds you with psychological safety in the midst of chaos is just one such positive example of an evolved limbic system.

💭 **Insight: Anchoring your emotional state to a positive memory helps regulate the amygdala, reduce stress hormones, and activate the parasympathetic nervous system.** Your brain *remembers* how to feel safe - if you remind it.

- Recalling one vivid, positive memory is enough to change your physiological state.
- Anchors work best when they're emotionally rich and sensory-specific and when you have a proverbial deck of them across the lifecycle.
- Repetition builds speed - the more you use your anchor, the faster it works.

This isn't about denial. It's about pause. It's about giving your nervous system a rope when the current starts to pull. Because life will knock you off center. That's a guarantee. But your anchor? That's your way back. An anchor that doesn't depend on another person or an object – an anchor that transforms your brain into an emotional touchstone library.

🧠 Future-Proof Your Brain Insight

Train your mind to return to calm - not through willpower, but through memory. In examining the limbic system and hippocampus as part of it, building dependence on your frontal cortex doesn't mean abandoning your limbic functions. It means having access to a foundation of choice in your reactions – rather than being a prisoner of mechanisms that no longer serves you. Picture a version of AI that you trusted to reference these gratitude anchors during AI therapy session when you were devolving into panic. The future of AI therapy includes precisely this level of personalization and trust. Until then, architect a portfolio of memories that trigger the feelings of gratitude and associated happiness. You cannot always control the external situations around you, but you can always create a mind that is built for cognitive longevity.

"I'll carry that moment with me. Not as a shield, but as a compass."

"Piglet noticed that even though he had a Very Small Heart, it could hold a rather large amount of Gratitude."
– A.A. Milne

Theme 5: Seasons, Environment, Evolution – Hello Neuroplasticity

Way 39: Habit Formation – but Make it Seasonal

Every January, millions of people declare their resolutions. "New year, new me." New gym pass. New planner. New pressure. And by February? Most of it disappears like bad snow, right it time for Valentine's Day resolutions about ideal relationships. 80% of New Year's Resolutions fail – a statistic that shocked the audience in January 2025 when I shared it on ***CTV News The Social*** in Toronto, Canada. The unpleasant truth is that January is one of the worst times to set goals. It's cold. It's dark – at least in much of North America.

Your energy is down. Your brain - and your body - are in conservation mode. But here's the secret: there *is* a better time to reset. It's Fall. The air shifts. The light softens. Even the trees let go of what no longer serves them. Long before smartphones and fitness challenges, we lived by the seasons. People used a Farmer's Almanac. And Fall signaled something important: It was time to re-evaluate, reset, and prepare. Not by force. By rhythm.

It's easy to forget that the current Gregorian calendar we operate by is only about 500 years old. Before that there was the Julian calendar, and even different rules for leap years. While the Gregorian calendar was adopted by most Catholic regions in the 1500s, a large group of countries didn't adopt it until the 18th or even 20th century. Now juxtapose those timelines with the human brain evolution timelines shared with you in Way 1, there's a clear dissonance.

The brain was built and went through massive inflection points of development and phylogenetic change at a time long before the current social calendar exists. There is a friction between our current calendar social expectations around goal setting and what the brain evolved to prioritize. The next time you feel the pressure to set New Years goals – remember this irrefutable fact: the brain wasn't built during the advent of the Gregorian calendar – it doesn't know that January means new resolutions, and most importantly, your brain doesn't care. It's also notable to be cognizant that in North America we often overlook that our seasons are not the seasons of the entire world. In Australia, for example, winter occurs from June to August – yes **JUNE**, a time when for North America we are ready for summer break to begin.

🧠 Insight: Fall offers the ideal psychological and environmental conditions for starting new habits.

- Natural transitions - like back-to-school, cooler weather, and shorter days - prime your brain to reflect and re-pattern.
- Unlike January, Fall isn't about pressure. It's about readiness.
- Your nervous system is looking for grounding - not reinvention.

Fall is the season of subtle transformation. It doesn't demand hustle. It *invites* reflection. It's a soft entry point into change. So instead of pushing your brain to change when it's least ready, align with the season that's quietly asking: What's worth keeping? What needs to be let go? What could begin again, gently?

🧠 Future-Proof Your Brain Insight

If you want sustainable change, sync with the world your brain evolved in - not the one your calendar sells. Try this thought on for size, like a perfect autumn knit sweater: *"I'll begin in Fall - because change works better when I don't force it."* Building a brain that understands the world doesn't stop because your mood plummets during rainy days is not mutually exclusive with planning the development of new habits during a season when they are most likely to stick with you, and you with them.

> *"Autumn is a second spring when every leaf is a flower."*
> *- Albert Camus*

Way 40: Pumpkin Spice and Everything Nice

There's a reason one whiff of cinnamon or roasted turkey can teleport you back in time. Or for vegetarians like me, the scent of pumpkin candles. Fall doesn't just look like change - it smells like it too. Warm spices, burning leaves, soup on the stove. These aren't just cozy clichés. They're **neural cues**. Your **olfactory center** is wired directly into your brain's memory hub - the hippocampus - and its emotion processor - the amygdala.

Unlike sound or sight, scent skips the line. It bypasses logic. And it brings feelings, images, and associations to the surface in an instant. Scent hijacks the brain because our olfactory system is hardwired to connect smells directly to emotional and survival centers, bypassing rational thought. Smell evolved before higher cognition - even bacteria "smelled" to survive. Bacteria can sense ammonia and other airborne volatile chemical compounds, which helps them locate "food" to survive.

Think: Thanksgiving. Or your grandmother's kitchen. Or the first day you wore the scarf she knit you and you didn't sweat. This is the season where memory and meaning merge, triggered by scent. So why not use that to your advantage?

💬 **Insight: Smell is the only sense with a direct line to the limbic system - the emotional and memory-processing core of your brain.**

- Pair new habits with distinct seasonal scents to strengthen memory encoding. If you're a TED Speaker, like me, I burn a pungent candle while I'm rehearsing any intense speeches – building stimuli complexity into my memory structure.
- Use aromatherapy intentionally - rosemary for focus, cinnamon for comfort, vanilla for calm
- Anchor routines to sensory rituals (e.g., same candle for journaling or reading)

Smell is more than atmosphere. It's *activation*. It tells your brain: "Remember this." And when you're forming new habits or letting go of old ones, anything that improves retention and emotional regulation helps.

So yes - light the pumpkin spice candle. Bake something nostalgic. Let your nose guide your memory. Let your memory guide your momentum. Aside from the delightful scents of pumpkins, vanilla, and pine – humans react intensely to pheromones. Pheromones are chemical signals/scents produced by humans and animals and signal a complex range of information: warnings of danger, nervousness, and the one we are most familiar with – attracting mates. Scent is just as imperative to cognitive upgrades as any of the other senses.

💭 Future-Proof Your Brain Insight
The fastest way to shift your mental state isn't always a thought. Sometimes it's a smell.

"I'll scent-mark my memories - and let Fall leave a trail I can follow."

"Perfume is like music that you wear"
– Mathilde Thomas.

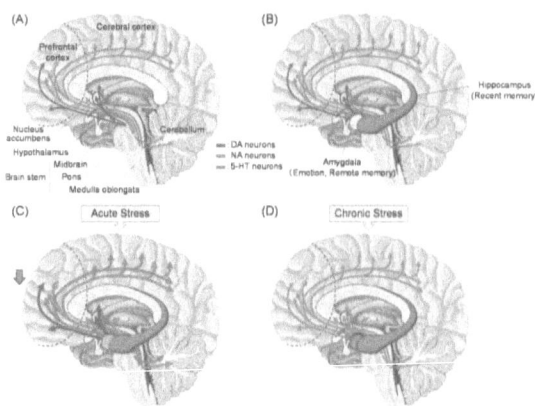

Image: Certain smells have stress-suppressing effects (Hussin, 2021), Pine or pinene is one of the most popular for stress-reduction

Way 41: The Snow is Coming

If you've ever felt like hibernating in December, you're not lazy. You're *aligned*. Your brain - and the environment it evolved in - isn't built for reinvention in Winter (no matter your cultural history). It's built for quiet. For depth. For pause. Anyone who's watched *Game of Thrones* knows: People in the North live differently than those in King's Landing. They prepare. They conserve. They endure. And it's not just because of White Walkers - it's because Winter demands a different rhythm. Long before artificial light and 24/7 calendars, Winter was our time to reflect, not reset. We gathered indoors. We told stories. We slept longer. We *listened* to the season instead of trying to outrun it.

My Canadian-ness is showing up yet again in these statements as snow was always part of my childhood – toboggans, sleds, skis, skates, and toques are ingrained components in all my nostalgia. But even for those in tropical climates, like my parents and grandparents who were born in Jamaica and Guyana, seasonal change triggers neurological change.

💭 **Insight: Reduced daylight in Winter increases melatonin and lowers serotonin - priming the brain for introspection and emotional depth.**

- Winter is ideal for deep focus, long-form thinking, and emotional integration
- Your energy is naturally lower - use it for recalibration, not overcommitment
- "Slowing down" is not a weakness - it's a **biological imperative.** And slowing down doesn't have to mean doing less. It can mean doing more activities that align to winter and its impact on your brain and metabolism.

This season isn't asking you to hustle. It's asking you to *digest* what the rest of the year stirred up. Winter is when the roots grow deeper. When the snow blankets distractions. When your inner world gets quieter - and richer. The intensity of how winter influences your brain is contingent on a number of genetic and environmental factors. Seasonal Affective Disorder (SAD) has been criticized by some scientists as unconfirmed, but seasonal depression undeniably exists. SAD has been found to be more prevalent at higher latitudes, i.e. northern geographies where sunlight exposure is reduced during winter (Chen, 2024). Lack of sunlight leads to reduced serotonin and melatonin, adversely impacting mood and sleep - treatment options range from bright white light therapy, Vitamin C in diet, to antidepressants.

Future-Proof Your Brain Insight

Listen to Winter. Let it be a signal - not of stagnation, but of deep internal change. Words of wisdom for you to repeat: *"I'll slow down with the season - because restoration is resistance."*

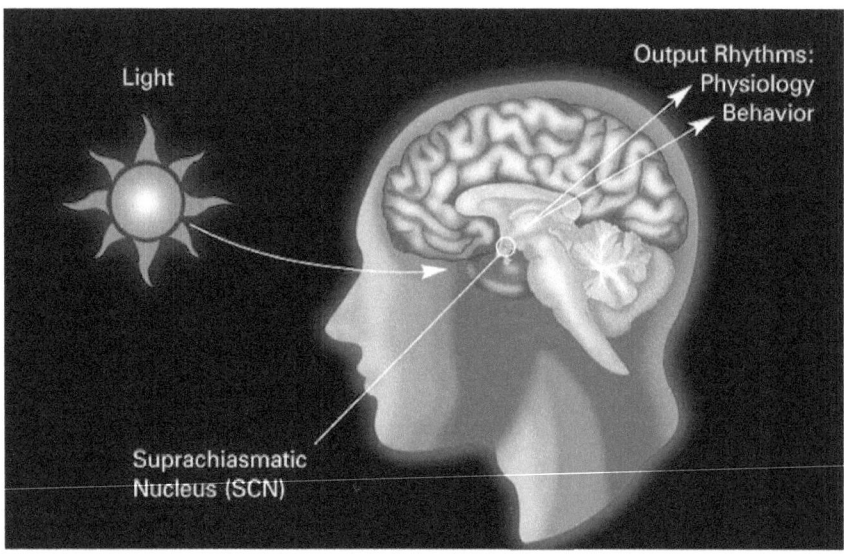

Image: National Institute of General Medical Sciences. Study from University of California San Diego – How your brain reacts to changes in the length of daylight.

Way 42: Coachella Summers and Beyond

There's a reason we crave music festivals, lake swims, and watermelon under the sun. Summer isn't just a season - it's a neurological invitation. School's out. The world opens to the possibility of play. Your body wants light. Your brain wants *space*. Those 8–10 weeks off for kids? That wasn't random. Historically, summer aligned with the harvest - a time to help on farms. But biologically, it also aligns with heightened openness, energy, and novelty-seeking behaviour.

Your brain is literally brighter in summer. Longer days increase serotonin. More light exposure improves mood regulation and emotional resilience. We're more social. More curious. More *ready to roam*. Those memes on social media about summer vacation in Europe versus summer in North America are funny, but also accurate – in Europe it's quite normal for people to take a month or longer off, and the practice makes a lot of sense in terms of neurological alignment.

Evolution occasionally gets a bad rap as a result of archaic research that erroneously categorized certain races as being at different points on the evolutionary curve, but the benefits of learning how and why we evolved in the trajectory we did can yield insights on how current day behaviours 1)can potentially help up continue to upgrade and evolve and 2) leverage "cheat codes" for neurological and lifestyle alignment.

💭 **Insight: Summer boosts dopamine and serotonin, making the brain more receptive to creativity, connection, and risk-taking.**
- It's the ideal time for brainstorming, collaboration, and creative play.
- Environments rich in sunlight and novelty support cognitive flexibility.
- Your brain's reward circuitry is more active - use it for exploration, not just indulgence.

This isn't the season for rules and rigid habits. It's for visioning. For loosening the grip. For asking: *What if...?* Let yourself be a little more impulsive. Let the conversations run late. Let the sun soften your urgency. You're not escaping. You're expanding.

🧠 Future-Proof Your Brain Insight

In Summer, your brain is wired for wonder. So, stretch into it - and let your neural pathways play. *"I'll think wide in the warmth - because every season grows a different part of me."* What might the world look like if we integrated seasons into work life instead of barreling through cognitive adjustments that result from seasonal changes? In Europe it's quite common to take 8 weeks of vacation for example. Even the 4-day work week proposed by some workplace consultants has merit – as people need at least 48-72 hours of complete cognitive reset to effectively extrapolate the benefits of weekend rest. As we consider the future of alternative work arrangements it's worth balancing the benefits of both ends of the spectrum – leveraging the positives of face-to-face interaction for social collaboration as well as realizing that beleaguered employees will not work harder simply because you reduce their vacation or personal time.

"Summertime is always the best of what might be."
– Charles Bowden

Way 43: Take a Hike…seriously.

That Chapter title sounds like a rude invitation, but it is a metaphor for diverting your brain to new neurological ventures. "Hike new trails." "Blaze new paths." Cue the inspirational quote on a mug. But here is the thing: hiking is not just poetic. It is tangible, clearly defined brain fuel. When you walk in nature - especially in unfamiliar places - your brain kicks into a whole new gear. You are navigating, adjusting, sensing, recalibrating. You are using spatial memory, proprioception, and real-time decision-making. And no, you do not have to befriend bears. Even a new city block or an unfamiliar park can unlock the same cognitive benefits.

If you have never been to Banff National Park, I highly recommend the experience in both summer and in winter – it is surrounded by what's called "The Valley of the Ten Peaks." On my first trip I hiked up the mountain in June 2018 and marvelled at the Kool-Aid turquoise colour of Lake Louise and that there was still snow on the ground in some places. I sat on a rock overlooking Moraine Lake and experienced the wonder of the Rocky Mountains.

On my second trip in November 2023, I proceeded to be equally shocked at how cold it was, the terror of avalanches and of getting stuck on the Icefields Parkway (yes, we got our car stuck - city people that we are). I experienced the wonder of hiking up to Ha Ling Peak and the pervasive feeling of stomping through untouched snow and being an "Explorer." The proprioceptive experience of covering an expanse stimulates the brain in ways that a treadmill or Peloton never can – even if you're wearing a virtual reality headset. Your innate operating system must take in a multitude of contextual data on a hike – wind, incline, vigilance against bears, colours, sights, smells, sounds, the difference in air quality. The treadmill in a gym cannot compare to that data input.

Insight: Novel physical movement in new environments increases brain-derived neurotrophic factor (BDNF), which supports memory, learning, and mood regulation.
- Hiking activates both the hippocampus (memory/navigation) and prefrontal cortex (decision-making)
- Nature exposure reduces cortisol and boosts creativity
- New terrain = new neural pathways - literally and figuratively

Spring is the season of emergence. The world thaws. The days stretch. Your brain, fresh from Winter's reflection, is ready to move - and map. Get outside. Take a different route. Let your body lead and your thoughts follow.

Long ago we were nomads where a sedentary life was antithetical to our continued survival. We knew how to navigate without Google Maps and how to tell time without Apple Watches. Our bodies yearned for new terrain. Technology continues to propel us forward to new discoveries at a fantastically exponential pace – but the complete disconnect with Circadian rhythms is undeniable. I was once asked if I think augmented or virtual reality will ever advance to the level of complexity where immersive realities can trick the brain – and my unapologetic answer as a consummate Futurist was: "I sincerely hope not!"

Future-Proof Your Brain Insight

To grow, your brain needs novelty *and* movement. Hiking gives you both - one step, one breath, one vista at a time. Visit places where your feet have never touched and your eyes are bombarded with landscapes that shock your brain into heightened sensory processing. Here's my hiking mantra: *"I'll walk new paths - not just to get somewhere, but to rewire the way I think."*

> *"In every walk with nature, one receives far more than he seeks."*
> *– John Muir*

Theme 6: Learning, Memory & Neurodiversity

Way 44: Hotdesking for Your Hippocampus

Switch it up. Change desks. Change seats. Change scenes. Not because you have to - but because your brain loves it. Sure, the whole return-to-office movement post-Covid 19 has been controversial. But hidden in the chaos of hotdesking and hybrid schedules? A neurological upgrade.

Every time you sit somewhere new - whether it's a new desk, a different café, or the other side of your home office - your brain reroutes. You're engaging your **parietal cortex**, responsible for spatial orientation, navigation, and mapping your surroundings. It's like giving your brain a mini map to redraw. And guess what happens when your brain redraws maps? It learns. Faster. Smarter. More flexibly. While employers are angling to increase productivity by maximizing supervision and in-person interaction – they're also limiting employees in terms of the impact that different work environments can have on increased creative solutioning.

In Way 35 I first touched on the insights of oxytocin and social bonding – the science is undeniable that in person interactions are fundamental to group dynamics, but this should not be mutually exclusive with acknowledging that not every employee thrives in a cubicle.

Insight: Changing work locations activates the hippocampus and parietal cortex - boosting memory formation, spatial reasoning, and mental adaptability.
- Novel environments stimulate neuroplasticity
- Switching locations even *once a month* can enhance focus and retention
- New surroundings = New synaptic patterns

Neurodivergent or neurotypical, your brain craves stimulation that's *just different enough*. So, if you can't escape the office, don't fret. Move desks. Claim a corner. Sit by a window. Or if you're home, rotate rooms. Work from the porch. New seat. New scenery. New circuits.

Future-Proof Your Brain Insight

Routine is easy. But reshuffling your space, even slightly, rewires your brain to stay curious and sharp. In Way 6 I shared the story of re-arranging my furniture during some of the most traumatic times in my life. We do not need to wait for a traumatic or destabilizing event in order to benefit from the outcomes of changing one's physical space frequently.

"I'll sit somewhere new - not because the view is better, but because my brain will be."

> *"I think the foremost quality-there's no success without it-is really loving what you do. If you love it, you do it well, and there's no success if you don't do well what you're working at."*
> *- Malcolm Forbes.*

Way 45: Draw-Storming

When words stop working, try lines. Shapes. Arrows. Doodles. Stick figures that don't make sense - until they do. In the unrefined scribbles sometimes, the brain finds itself unshackled and able to discover to insights. Be silly and create images that don't have to be perfect or mean anything. I remember doing this all throughout elementary school and high school – doodling in the margins of my workbooks when I felt stuck on writing an essay or solving a combinatorial math problem. There were times when I would write my name in strange made-up fonts until it looked unfamiliar even to me. But once I went to university and I became shackled to my laptop, doodling fell off my radar, and I rarely took notes on paper. In Way 33, I elucidate the concept of "going cold turkey with tech" and all the ways in which digital detox can be just as beneficial and digital upskilling (we really do need both). "Draw-storming" is a great way to fill that space left by reduced digital interaction.

We've all been told to brainstorm. But sometimes, the storm needs a different kind of spark. Enter: the pencil. Or pen. Or crayon. Or finger on a touchscreen. When you draw - especially when you're stuck - something strange and wonderful happens. You stop explaining. You start seeing. That's your **occipital lobe** lighting up. It's your visual cortex doing what it does best - making sense of the abstract by drawing it into form. Ask yourself, when is the last time you doodled or coloured as an adult? For those of us who aren't artists, just how often do you utilize the part of your brain that imagines the liminal, the amorphous, and the microcosms of labyrinths – is there space for imagination without any purpose?

🧠 **Insight: Visual sketching engages the occipital lobe and boosts connectivity between the brain's visual and associative areas, improving ideation, memory, and insight.**

- Drawing bypasses language blocks and taps into visual-spatial reasoning
- Even crude sketches reduce mental friction during problem-solving
- "Drawing your thoughts" helps clarify emotions, plans, and ideas

You don't need to be an artist. You just need to let your pen move before your inner critic can stop it. So next time you're stuck - don't write.

Draw. Map your thoughts. Sketch your frustration. Doodle your way to the answer.

🧠 Future-Proof Your Brain Insight

When you move from words to visuals, you unlock different neural highways - ones built for pattern recognition, clarity, and creativity. Try to prioritize actual paper and pen/pencil scribbling – the sensory feedback from an iPad cannot replicate what your brain records when it processes the feeling of your writing instrument touching the texture of a page. *"When I can't think straight, I draw crooked."*

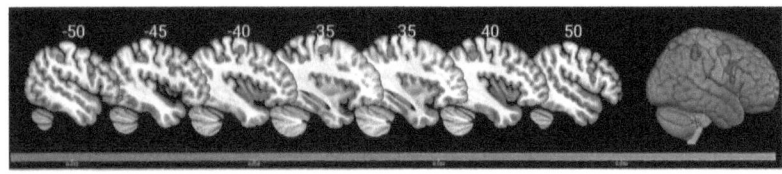

Image: The Likelihood of Neural Activation During Drawing (Raimo, 2021)

Way 46: Around the Campfire ?

Long before phones and feeds, we gathered around fires. Not just to keep warm, but to remember - and to be remembered. Storytelling was how we made sense of the world. It helped us connect dots across time, explain the unexplainable, and pass on the lessons we couldn't afford to lose. But most importantly, it gave meaning to *experience* - by turning moments into memory.

Today, we still tell stories. But they often come in quick hits: a three-second story on Instagram, a message peppered with emojis, or a meme that stands in for our mood. Various DMs of jokes and videos we send to each other – each easily forgotten after a quick giggle or forwarding it onto someone else.

When was the last time you *really* told a story - start to finish - with your own voice, your own pacing, your own perspective? When you take time to narrate your own experiences, your episodic memory kicks in. This is the part of your brain responsible for recalling events in context - where you were, how you felt, what it meant. And the more you engage with those memories actively, the more vividly they're encoded and retained. Research also reveals that naming emotions and feelings can mitigate the potentially negative effects of those feelings (Kircanski, 2012). Storytelling is the lost art of demonstrating to each other our unique thought processes outside of the confines of Outlook and Google e-mails.

🧠 **Insight: Storytelling strengthens episodic memory by engaging the hippocampus and default mode network - the brain's "autobiography mode."** Narrating experiences activates deep memory and emotional understanding

- Telling a story - out loud or in writing - helps the brain reconstruct and rewire events with greater clarity
- Sharing stories in different formats (voice, writing, conversation) diversifies your neural engagement

Start small. Send a voice note instead of a text. Write an email that doesn't just summarize - but tells the story behind the summary. Talk over dinner about something funny, or painful, or real that happened to you this week. Tell the full story. Not because someone else needs to hear it, but because your brain does.

🧠 Future-Proof Your Brain Insight

To strengthen your memory, don't just collect experiences - recount them. And the anecdotes need not only be positive with happy endings. The best fairytales all need some adventure. Storytelling was the mechanism for passing on knowledge for many a millennium – one that still captivates human audiences in the modern day. *"I tell stories not to be heard, but to remember who I am - and who I've been becoming."*

> **"There is no greater agony than bearing an untold story inside of you."**
> *– Maya Angelou*

Way 47: Space – the Final Frontier?

No, not that kind of space (proud Trekkie here!). Not warp drives or distant galaxies, or Klingons, or Captain Kirk We're talking about mental space - the kind your brain needs to learn, remember, and perform. In 2023 when I finally took the stage to deliver my very first TED talk in Toronto, I had 17 years of professional speaking experience. But for keynotes a lot of the performance is open to improvising on stage, reading the room, and selecting anecdotes that most resonate with how people are reacting. When I gave my TED talk, the guidelines were intimidating:
- No confidence monitors
- No teleprompter
- 100% script of your talk memorized verbatim
- With a small variance allowed by the organizers of 5% (otherwise? They wouldn't be submitting your talk to TED for publication).

How was I going to do this? In my life sciences work I had always prided myself on an eidetic memory, but how was I going to merge my improvisational public speaking style of more than a decade with memorization?? Since my talk touched on themes of neuroscience, I turned to the science of memorization and spaced repetition.

Spaced repetition isn't a hack. It's the way your brain *wants* to remember - not through cramming, but by returning again and again to what matters. But here's the catch: spacing alone won't save you if what you're reviewing is chaotic, unhinged, and poorly written. Organization and structure paradoxically create space for your brain to be creative with memory coding.

Enter chunking - the second half of the memory equation. Chunking is how we organize information into meaning. It's why phone numbers are grouped in threes and fours, why speeches fall into beats, and why actors break scripts into emotional moments, not paragraphs. Miller's Magic Number is the theory developed by psychologist George A Miller in 1956 that says the ideal number of items we can concurrently remember is 7, plus or minus 2. This aligns with many examples we see in real life – like phone numbers. Want to remember more with less stress? Chunk first. Then space it out.

🧠 Insight: Chunking + Spaced Repetition = Memory Magic

- **Chunking reduces cognitive load.** Your brain can only hold 4–7 things at once, or 5-9 if you push it's limits. Grouping info into meaningful "chunks" makes it easier to learn and rehearse.//
- **Spaced repetition strengthens the signal.** When you review each chunk at increasing intervals, the memory trace becomes sharper, stronger, and more permanent.
- **Together, they scaffold long-term memory.** First, create structure with chunks. Then, let time and spacing reinforce each one - like layering bricks in a mental wall. You may believe cramming for exams or work presentations works – but it only works if it is bolstered by functional understanding.

Real-Life Example: Memorizing a Script

1. **Break it into beats or emotional shifts.** Don't memorize page-by-page. Group lines by meaning. Try to architect a story in your mind.
2. **Assign gestures or tones to each chunk.** This anchors the memory in physical and emotional cues. When you see public speakers "hand waving" – aside from the fact that this sends a social signal to the audience to engage, it's also part of the craft for many performers in helping them to remember. This is the intersection of physical motion x verbal cues.
3. **Review each chunk separately - then together.** Space out your review: 10 minutes later, then 1 hour, then the next day. I recited my TED talk more than 170 times…and yes, I badgered my boyfriend (now my husband) into being the mock audience! This method isn't just for actors. Use it for speeches, pitches, keynotes - even exam prep. Your brain doesn't want to just store data. It wants to tell stories and recognize patterns. *See Way 84 for more on Storytelling

🧠 Future-Proof Your Brain Insight

To encode information deeply, don't just space it out - give it shape first. Chunking gives structure. Spaced repetition gives strength. You remember more when meaning is made - and revisited. A great example is learning to drive – you learn how to parallel park by chunking together the instructions: pull alongside the car in front of the space you intend to park in, begin angling your wheel when the front of your car is at about the halfway mark of the car parallel to you, once you are at a 45 degree angle then begin straightening up your wheel and reverse neatly into the space, all while checking your mirrors and the space behind you.

This is a lot of information for your brain to process: verbal instructions (possibly from one of your parents yelling at you as you learned to drive), visual data from at least 3 mirrors, possibly your cars proximity sensors beeping, and often if you're a woman, like me, the fascinating stares of people when you parallel park without assistance. It's truly a miracle we learn to drive at all when you think of how much information we have to concurrently keep track of – but chunking it together is precisely what saves us!

Image Source: Start Rescue U.K.
"Parallel Parking in 6 Steps"

Way 48: The Castle

In today's age of constant change, it can feel overwhelming to remember information from every single webinar and training session you feel you're being subjected to or you've optimistically registered for. Your brain and its memory tools are one of the most important lifelines you have in constant change. Some memories are like keys. Others? Castles. Built room by room, each hallway leading to another idea, each door marked with meaning. These memories exist as an interconnect scaffold.

If you want to remember more - and remember it deeply, don't just study harder. Build it. Visually. Spatially. Personally. This is the method of memory palaces - an ancient tool for a modern mind. The Greeks used it to remember epic speeches. Today, actors, surgeons, lawyers, and memory champions use it to remember complex scripts, sequences, and case law. You don't need perfect recall. You need a *place* to put what matters. A mental house built of ideas, anchored in space and story.

Insight: Why Memory Palaces Work
- **They activate the hippocampus - the brain's spatial and episodic memory hub.** When you visualize a physical place and walk through it mentally, your brain encodes the journey like it's real.
- **They leverage the brain's preference for imagery.** Visual memories stick better than abstract ones. Add **colour, motion, exaggeration**, and emotion to lock it in.
- **They connect concepts into a schema.** Instead of isolated facts, you create an interlinked system. The more fulsome and layered the visual, the more robust the memory.

Try This: Build Your Castle
1. **Pick a place you know well.** Your childhood home. A favorite café. A gallery you wandered once. The more emotionally vivid, the better.
2. **Assign an idea to each room.** A theory in the kitchen. A hypothesis in the hallway. A concept visualized as a painting in the den.
3. **Walk through it. Pause. Touch. Feel.** Add exaggerated imagery. Let a concept explode into fireworks. Place a model of the solar system in your shower. Make it unforgettable.

4. **Return to your castle regularly.** Like all memory work, repetition cements the structure. Update it. Expand it. Build a new wing when you learn something new.

This is beyond a vision board. This is embodied cognition - the mind remembering what the body imagines. It's the connection of all your senses.

🧠 Future-Proof Your Brain Insight

Don't just store information. **Stage it.** Give your ideas space, visuals, texture - and they'll never leave you. This provides an incentive for your brain to recollect information more effectively – because returning to a beautiful castle is much more appealing than a blank room. *"The best ideas live in rooms you can return to."*

Image: Palacio de Pena, Sintra, Portugal – one of my memory castles for experiences that signal success

Theme 7: Your Body vs. Brain Health

Way 49: Alpha & Omega

You don't need to overhaul your entire life to future-proof your brain. But you do need to start at the source. With what you eat. With what you absorb. With what your brain is literally made of. Because your thoughts? They're built on lipids. And your memory? It's wrapped in fat - quite literally.

That's where **Omega-3 fatty acids** come in. They're the building blocks of your brain's structure and function - crucial for maintaining cell membranes, reducing inflammation, and supporting everything from mood regulation to long-term memory. As a vegetarian for 27 years, healthy fat in diet can be a challenge – especially with a food sensitivity to FODMAP foods. FODMAP stands for **fermentable oligosaccharides, disaccharides, monosaccharides and polyols –** and yes, those delicious avocados fall into this category. I do so dearly love guacamole...but it doesn't always love me.

🧠 Insight: **Omega-3s and Brain Health**
- **Your brain is 60% fat.** Omega-3s, especially DHA and EPA, are essential to neuron structure and synaptic communication.
- **They modulate inflammation.** Chronic, low-grade inflammation is a silent disruptor of cognitive function. Omega-3s counteract that.
- **They influence neurotransmitters.** Higher Omega-3 intake has been linked to improved serotonin and dopamine regulation - essential for mood and mental clarity.

Vegetarian or vegan?
You're not out of luck. You're just shopping in a different aisle.
- **Algae oil** is a plant-based powerhouse - and one of the only vegan sources of DHA and EPA.
- **Flax, chia, hemp, and walnuts** contain ALA, a precursor to DHA.
- **Supplement smartly.** Many vegan Omega-3 blends now deliver bioavailable forms without fish or gelatin.

Pea protein is particularly helpful if you are part of the cohort like myself who doesn't consumer animal-based proteins (no judgement on those you do!). Pea protein contains palmitoylethanolamide (PEA) (a mouthful and serendipitous acronym), which reduced brain fog and combats neurodegeneration (Na et al. 2021). PEA also combats chronic pain and has numerous neuroprotective properties.

Bonus Brain Tip: Find Your Nutrition Person

Neuroscience doesn't end at the MRI machine. It runs through your kitchen, your supplements, your sleep. Work with someone who gets the big picture - not just calories, but cognition. Recently, I received a tough health diagnosis that can be mitigated by dietary adjustments. I decided to eschew and delay invasive surgery for as long as possible. As I was researching solutions for anti-inflammatory diets I had conversations with nutritionists, doctors and naturopaths. As a neuroscientist I love the hard factual evidence I get from an MD, but building a fulsome health solution drove me to consider numerous credentialled perspectives. Conversing with AI should never be a replacement for a qualified medical doctor, but it can open your mind to building a health plan that incorporates multiple modes of treatment. A recurring theme in health trends is limited focus – diet, exercise, rest – but all roads lead back to the brain. It is the synthesizer of everything you process and consume.

🧠 Future-Proof Your Brain Insight

You are what your brain eats. Feed it **omega-rich, anti-inflammatory fuel** - and give it the support team it deserves.

"Neurons don't just fire on willpower. They fire on what you fed them."

"There are only two sorts of doctors: those who practice with their brains, and those who practice with their tongues."
-   William Osler

Way 50: Walking….Really it's that simple

We think of walking as basic. Ordinary. Automatic. Perfunctory. Counting your 10,000 steps on your Apple Watch sounds anything but novel. Every step you take - post-dinner, mid-phone call, or during a quiet thinking stroll - is activating ancient circuitry your brain still depends on today. Walking isn't just movement. It's mental calibration. When you walk, several powerful neurological systems come online (Wojtys, 2015)– ones that are older than your prefrontal cortex.

🧠 Spinal Central Pattern Generators (CPGs)

These are rhythm generators embedded in your spinal cord. They produce the patterned signals that drive your legs - **even without your brain telling them to.**

It's your nervous system on autopilot, evolved to keep you moving forward.

🧠 Cerebellum & Brainstem

These regions coordinate balance and precision. Without them, steps become stumbles. Their role is quiet - but essential. Every smooth pivot, every mid-step adjustment? That's them, tuning your trajectory in real time.

🧠 Sensory Feedback Loops

As you walk, your joints, feet, and muscles talk back to your brain. Tiny corrections. Micro-calibrations. This feedback keeps your gait stable and aligned - especially over time and terrain. And beyond locomotion, walking activates another powerful region:

🧠 The Hippocampus

The same structure responsible for forming new memories lights up after a short walk. Especially after eating - when blood flow and glucose are both on the move.

Walking after a meal? Old-school advice - yes. But now backed by modern imaging. So no, you don't need to meditate on a mountain. Or buy a Peloton. You just need to move your legs and let your body do what it was designed to do. A 40-minute low-intensity walk can result in significant changes, especially if it is outside as opposed to a treadmill. (McDonnell, 2024).

🧠 Future-Proof Your Brain Insight

Walking isn't just exercise - it's **neurological maintenance.** When your grandparents would try to get you to take a walk after dinner, they knew what they were talking about. Rhythm. Balance. Memory. Coordination. Step by step. *"Some of your smartest thoughts will arrive on foot."*

Image: The Nike "Mind" Shoes, created with the intention of giving more sensory feedback to the brain.

The creators of the shoe shared publicly in January 2026 that in a synthetic environment wearers could "feel" the artificial blades of grass in a lab as they walked. Is this "neuroscience shoe"? That's still up for debate. The increased physical stimuli might just be more of a distraction than a help to wearers.

A visual of the massive amount of neural activation that occurs from the simple act of walking:

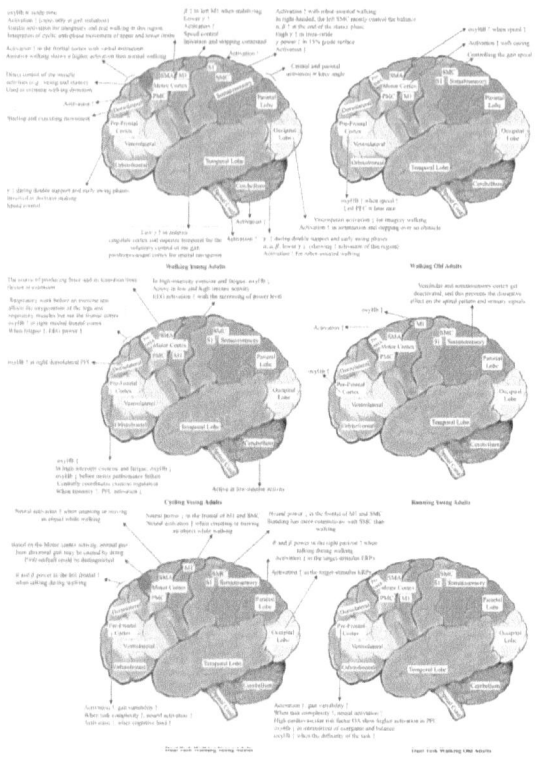

Image: The Impact of Locomotion on the Brain (Korivand, 2023).

"All great thoughts are perceived by walking."
- *Fredrich Nietzche*

Way 51: HiiT

We don't usually associate gasping for breath with better memory or any positive emotions. But science says otherwise. That moment when your legs burn, your heart pounds, and your lungs feel like they're on fire - your brain is listening - and it's changing. We all know exercise is good for us, but we're not often inclined to push our bodies to the breaking point.

Enter **HiiT** - High-Intensity Interval Training. Short bursts. Full throttle. Rest. Repeat. And the impact on your brain may last years. According to a 2023 study published by *Harvard Health*, participants who engaged in high-intensity training had increased levels of **BDNF - Brain-Derived Neurotrophic Factor** - a key protein responsible for neuroplasticity, learning, and memory. And here's the kicker: That spike in BDNF wasn't just short-term. The cognitive benefits lasted for years after consistent HiiT.

🧠 Insight: Why HiiT Supercharges Your Brain

- **BDNF is like Miracle-Gro for your neurons.** It supports the growth of new brain cells and strengthens synaptic connections.
- **High-intensity training increases BDNF more than moderate exercise.** Sprinting, biking, jump squats - it's the intensity that unlocks the neurochemical flood.
- **The hippocampus thrives under pressure.** This region, crucial for learning and memory, lights up during and after high-intensity bursts.
- **Long-term impact:** The study tracked older adults and found that cognitive function remained sharper even one year after the HiiT interventions ended.

How to HiiT It Right

You don't need fancy gear. Just intervals. Try this:
1. Warm up for 5 minutes.
2. Sprint or cycle hard for 30 seconds.
3. Recover for 90 seconds.
4. Repeat 4–6 times.

That's one brain-boosting session in under 20 minutes.

Future-Proof Your Brain Insight

If you want to grow your brain - not just your biceps - push yourself into short bursts of intensity. Your neurons will thank you. Endurance, like neurons, doesn't grow unless it's pushed to the limit. An unpopular lesson from neuroscience is that pushing the boundaries of exhaustion and friction within the brain can yield stronger neural synapses.

> *"Neuroplasticity doesn't wait for a perfect plan. It responds to effort."*
> - *Sarah Baldeo*

Way 52: It's all about Balance

I try to remind myself daily about balancing life and work – rarely is there ever a balance. One practice that I've been able to maintain is 20+ years of doing yoga. Balance isn't just about standing on one leg. Or flowing through Ashtanga yoga poses. It's about neurological coordination - the intricate dance between your body and your brain. Proprioception is the way we perceive our body in relation to others and our surroundings, it's how you know how far it is to shake someone's hand. Interoception is our awareness of our own internal mechanisms – like the beating of our heart when we're anxious about a job interview.

Balance can merge both proprioception and interoception. And the payoff? Not just better posture. But better cognition, memory, and spatial awareness. Your brain doesn't treat balance like a background task. It throws major resources at it - activating your **vestibular system**, **cerebellum**, **prefrontal cortex**, and **hippocampus** in a single wobble. Which means: When you train your balance, you train your brain.

💭 Insight: Balance is Cognitive Training in Disguise

A 2020 peer-reviewed study published in *Frontiers in Human Neuroscience* (Rogge, 2020) found that 12 weeks of balance training in older adults led to measurable improvements in brain structure and function - specifically in areas tied to memory, attention, and sensorimotor integration.

Here's why it works:

- **Vestibular Activation:** Your inner ear system constantly adjusts to shifts in position, posture, and motion - feeding your brain real-time updates to keep you upright. This stimulates the hippocampus, crucial for both navigation and memory.
- **Neuroplasticity:** The act of keeping your balance engages multiple brain regions at once - creating new connections, refining motor control, and enhancing adaptability.
- **Cognitive Load:** Dynamic balance tasks demand focus, problem-solving, and adaptability - all of which sharpen executive function.

Try This
- Stand on one foot while brushing your teeth.
- Use a balance board while reading emails.
- Try walking heel-to-toe in a straight line with eyes closed.

- Or go for a modern twist: slackline, dance, martial arts, tai chi.

The goal isn't stillness - It's **dynamic instability**. Controlled chaos. That's where the growth happens.

🧠 Future-Proof Your Brain Insight

Balance isn't just physical - it's **cognitive scaffolding**. Train it, and you train your brain to adapt, anticipate, and stay sharp. Expecting constant balance even beyond the physical to work-life balance, life-work balance, is a myth. The world and deadlines do not operate on a trolley cart of your emotions – there will be times where an upgraded brain means compartmentalizing transient and temporary emotional upheaval. *"The more you challenge your balance, the less life throws you off center."*

"Next to love, balance is the most important thing."
  - ***John Wooden***

Way 53: Throw a Dance Party!

Whether you're a classically trained dancer or someone who confuses samba with spaghetti, dance is one of the best things you can do for your brain. Yes, even if you're not good at it!

Not your vibe? That's okay. Because this isn't about performance. It's about **neural fireworks**. Dancing taps into a unique brain-state - where movement meets music, memory meets rhythm, and emotion meets coordination. It's a full-brain workout. And it's backed by science.

Insight: Dance Lights Up Your Brain

- **Motor Cortex:** Every spin, step, and sway activates your motor planning regions; improving coordination, flexibility, and proprioception.
- **Auditory Cortex:** Following the beat? That's your auditory brain at work - synchronizing sound and movement in real time. Even if you have "no rhythm," just your attempts to synchronize prompts increased neural activity.
- **Memory Systems:** Learning choreography? Even casually? That's procedural memory. You're strengthening your **hippocampus**, your **basal ganglia**, and even your **prefrontal cortex**.
- **Emotional Regulation:** Movement + music = dopamine, endorphins, and a major drop in cortisol. Translation? Dancing helps you feel good, think clearly, and bounce back faster.

Research has shown that regular dance sessions can improve executive function, spatial awareness, working memory, and even help stave off cognitive decline in aging adults. It's no wonder dance therapy is now used in neurorehabilitation clinics around the world.

Try This
- Dance in your living room once a week.
- Learn one new move every Friday.
- Revisit the music you grew up with - and let your body remember.
- Or go all in: bachata classes, hip hop drop-ins, salsa under the stars. You don't have to be good. You just have to move to music.

🧠 Future-Proof Your Brain Insight

Dance isn't just fun - it's a neurological symphony. When you move to music, you're synchronizing systems across your brain and body. And that kind of harmony? It lasts. *"Dance like no one is watching - because your brain always is."*

> *"I always work hard to find a way to disconnect from the thinking until it becomes second nature to me because that's where you find the best moments. Dancing is like that for me all the time. It makes me feel free."*
> — *J Lo.*

Theme 8: Relationships & Tribes

Way 54: Explore outside Your Tribe

As kids, we were curious about everyone. The new kid. The old neighbor. The cousin with a strange accent. But somewhere along the way, we built circles - tight-knit, familiar, echoing back our own beliefs. And while those circles offer comfort, they can also create cognitive stagnation. If you want a more flexible, future-proof brain, you need to step outside them - regularly. When I was an Executive Director at one of the biggest consulting firms in the world I was told frequently, "you just need to find your tribe!" But I didn't want to belong to just one tribe, I wanted to enjoy connection with many different types of leaders!

Once a month, I commit to having coffee with someone different than myself in every way possible. In person or virtual. Across cultures. Across industries. Across life stages. In my family there are about 9 different cultures - Irish, German, Spanish, French, Chinese, West Indian, Russian, and Italian. The idea of "sticking to my own kind" never resonated with me. Within my friend circle there are numerous opinions on interracial data. I seek out diversity in connection not just because it's polite or DEI friendly. Because it's **neuroplastic**. And perhaps most crucially, a lack of alternate perspectives invariably leads to kakistocracy. If you haven't heard of this word, you'll love using it. It means a government or body of authority that is run by the most incompetent leaders.

🧠 Insight: The Neuroscience of Diverse Perspectives

- **Neuroplasticity thrives on novelty.** New conversations challenge your predictive brain. They break mental routines and force fresh associations.
- **Perspective-taking strengthens the prefrontal cortex.** This is the part of your brain responsible for empathy, self-regulation, and complex decision-making.
- **Social discomfort activates growth.** When you encounter unfamiliar beliefs or lifestyles, your brain adapts - by rewiring old judgments and forming new cognitive pathways.

Try This

- Grab coffee with someone 20 years older or younger than you.

- Ask a parent what they've learned from parenting - even if you're child-free.
- Meet a researcher if you're a creative, or a tradesperson if you're in tech.
- Listen more than you speak.
- Ask each of these people them what they believe - and **why**.
- Find a quiet space – take a pen and paper and write a page about yourself. Imagine you're at an event (not for work). Try to describe yourself without referencing your gender, age or ethnicity – and don't even think about including what you do for work. Who are you beyond the elements that you don't have any part in choosing? Who are you beyond what you do to generate income? The goal isn't to agree. It's to grow.

Future-Proof Your Brain Insight

Familiarity is easy. But diversity builds depth. To stay sharp, curious, and emotionally agile - stretch your social network beyond your defaults. Audit your friend circle and confront the potentially uncomfortable reality that all your friends aren't that much different than yourself. *"Your brain doesn't grow from being right. It grows from being surprised."*

"It is time for parents to teach young people early on that in diversity there is beauty and there is strength."
- Maya Angelou

Way 55: Go Clubbing...

Okay - the title's a little misleading. But hear me out. This isn't about sticky dance floors and over-priced bottle service. This is about joining clubs, hobbies, spaces that have nothing to do with your job. Because your brain? It wasn't built to think about work all the time. It was built to wander, tinker, try things, and play. Anyone who knows me will marvel at the fact that I'm saying this, because I really love to work and be productive.

Most of us - especially the driven ones - join professional boards, volunteer for causes that overlap with our careers, and call that a "break." I'm the first to admit I'm guilty of that – every board or group I joined over my career has been tired to my work. Spoiler: that's not "a break." Your brain craves novelty and non-utility. You don't always need a return on investment. Sometimes, what your neurons need is something pointless... and joyful.

Recently in one of my life coaching sessions a client mentioned to me that she was struggling with her husband returning home from active duty. She'd gotten so used to him being away and adjusting to having him home and trying to relearn a shared family dynamic, was "driving her nuts." So, we built a plan for her to experiment with knitting – something she had never done and had zero interest in doing ever. This activity was so vastly divergent from anything she'd ever done before – it acted as a pathway to helping her brain be primed for adjustment.

Insight: Why Hobbies Matter for Brain Health
- **They activate default mode and creative networks.** Hobbies encourage open-ended thinking - the kind that leads to insight, imagination, and emotional release.
- **They reduce cognitive rigidity.** When everything you do is tied to expertise, you stop taking risks. Hobbies make you a beginner again - and that rewires your brain in all the right ways.
- **They lower stress and increase longevity.** Studies have shown that people with strong hobby engagement have lower cortisol, improved sleep, and even longer life spans.

Try This
- Join a birdwatching group.

- Take improv or pottery or fencing.
- Learn to DJ.
- Build a tiny model village (you're an adult, you can do whatever you want).
- Or just join a club that meets once a month and talks about anything *but* work.

It doesn't have to be "productive." It just has to be an activity that you do outside of work.

Future-Proof Your Brain Insight

Don't let your career consume your curiosity. The brain is like a muscle - and cross-training counts. You're not just what you do. You're what you love when no one's watching. Investing in activities that are not tied to revenue generation or social status create a space within your brain that doesn't have to map to metrics of success in a social hierarchy.

> *"Hobbies are great distractions from the worries and troubles that plague daily living."*
> *– Bill Malone*

Way 56: The Imitation Situation

We talk about being a "role model" like it's something for kids. But what if the real secret to growth - even as an adult - is watching others more closely? A fact that seems bizarre to most is that your brain is wired to imitate. To observe, internalize, and adapt the behaviours of people around you. And this isn't just social advice. It's neuroscience. At the center of this process are your mirror neurons - specialized brain cells that fire not only when *you* do something, but also when you watch someone else do it.

Whether it's kindness, calm under pressure, or the way someone solves problems - when you see it, your brain rehearses it. Over the years I've had the privilege of travelling all over the world to connect with audiences during my keynotes. On a recent trip to Kuala Lumpur, I met with a fascinating executive leader at an HR conference. She was originally from the UK but had moved to Malaysia for work. After 10 years of living in Malaysia, she'd noticed that her accent began to change – she was inadvertently copying the intonation and inflection of the people around her. In the same way we can unconsciously begin to internalize accepts, we easily absorb behaviours and belief systems like giant adult sponges.

Insight: Learning by Observation Changes Your Brain

A 2014 study published in *Frontiers in Human Neuroscience* (Iacoboni, 2014) explains that mirror neurons play a key role in social learning, empathy, and imitation - especially in contexts where a person strives to improve or acquire new behaviours.

Here's what that means:

- **Mirror neurons help encode observed actions as if you performed them.** Simply watching someone handle stress with grace strengthens your own potential to do the same.
- **These neurons connect empathy and learning.** You don't just copy behaviour - you absorb its context, its tone, its emotional undercurrent.
- **They stay active throughout life.** You never "age out" of this learning system. But you do have to use it - intentionally. If you're unwilling to adjust your ingrained patterns, it doesn't matter how much you observe.

Try This
- Next time you admire someone's way of handling conflict, take mental notes.
- If a colleague has a habit you respect - punctuality, presence, patience - ask how they built it.
- Mirror positive behaviour in small doses. Try it on. Modify it. Make it yours. Straight mimicry isn't the goal – it's emulating and personalizing behaviours that's the winning combo. Growth isn't always about self-generated insight. Sometimes, it's about **selective imitation** - done consciously and with humility. After all, we can all learn from others, even (sometimes especially) as adults.

Future-Proof Your Brain Insight

You're never too old to learn by watching. Mirror the traits you admire - and your brain will do the rest. *"We become the people we observe - especially when we choose them wisely."*

Way 57: Mentor

We usually think of a mentor as someone older, wiser, and farther along the path than we are. But what if mentorship isn't about age or titles - but about shared growth? After 2 years as a Mentor with Women's Infrastructure Network I realized more than once that I was learning so much from my mentee. At Deloitte as part of the Mentors Program partnership with University of Toronto, the fresh perspectives of my mentees drove away my cynicism more than once. And conversely, I learned so much from my mentees. Mentorship works both ways. Your brain is wired not just to teach, but to thrive through connection - especially as you age. In fact, mentoring itself might be one of the most powerful tools to keep your mind sharp and your sense of purpose alive.

Mentorship when done right, can eradicate some of the erroneous and potentially harmful stereotypes about age and generations. I happen to be a millennial, but I have met individuals both younger and older than I who have equally upgraded my knowledge. As pedagogical approaches continue to change and education evolves, age is no longer the sole determinant of experience and wisdom.

🧠 Insight: Mentorship Strengthens Your Brain

A 2013 neuroimaging study (Brown, 2013) found that midlife and older women engaging in caregiving and mentoring behaviours show greater activation in brain regions related to empathy, social cognition, and reward - suggesting the brain is primed to find meaning in mentorship as we age.

Here's what that means:

- **Mentoring others activates empathy, perspective-taking, and purpose circuits.** It's not just altruistic - it's neurologically nourishing.
- **It's reciprocal.** You teach, but you also absorb - through listening, reframing, and reflecting.
- **It's cross-generational.** The brain doesn't care if you're mentoring up or down in age. The growth happens either way.

Try This

- Seek out someone **10 years younger** than you - offer your time, stories, and lessons.

- Then find someone **10 years older** - and ask about their decisions, their regrets, their joy.
- Approach both relationships as mutual learning experiences, not lectures or hierarchies.

🧠 Future-Proof Your Brain Insight

Mentorship isn't about being the expert. It's about being **engaged** - with people, with ideas, and with growth across generations. You'll stretch your empathy, expand your perspective, and rewire your brain - one conversation at a time. The best mentors never stop learning. And the best students never stop teaching.

"Age is not a barrier. It's a limitation you put on your mind."
- Jackie Joyner-Kersee

Way 58: The Mirror Project

In Way 35 and 38 gratitude was a popular character that showed up more than once. And in Way 56 mirror neurons entered the group chat. But mirror neurons are not only useful in teaching positive behaviours to your social circle or children. Blending the science of gratitude into the mix is about socially conditioning yourself and the people around you to be kinder.

You've heard of gratitude journals. You've probably heard of random acts of kindness. But what if the next level of growth isn't about *feeling* grateful - it's about reflecting it back? Appreciation is contagious - but only if it's expressed. Your brain is wired to mirror the emotional tone of the people around you. When someone shows you genuine appreciation, it does more than warm your heart - it lights up your mirror neurons, the same way it would if *you* had said or done the thing.

This isn't just about being nice. It's about building an ecosystem of encouragement - one that rewires both your brain and the brains of those around you. Years ago, when I worked at Ceridian before it became Dayforce we would write hand-written thank you notes to clients for Christmas – sending e-cards was considered lazy. In an average year each member of the business consulting team may very well have worked with 1000s of business owners – but the practice of hand-writing notes made a huge difference in how we all remembered each client and how they felt we viewed them as more than just a transaction.

💬 Insight: Gratitude and Mimicry Are Linked by Design

Neuroscientists have long studied how **mirror neurons** support learning through imitation, but recent research also suggests they play a role in **emotional contagion** - the unconscious mimicry of positive (or negative) emotional cues. When you express sincere gratitude, you prime the other person to do the same - not just behaviourally, but neurologically.

Here's what that means:
- **Gratitude isn't just felt. It's modeled.** People around you learn how to appreciate by watching how you do it.
- **Mirror neurons amplify behaviour.** Your expressions - kind or critical - are neurochemical cues others may unconsciously adopt.

- **Children aren't the only mimics.** Adults mirror too. Often subtly. Often habitually. Gratitude, when modeled authentically, spreads.

Try This
- Start your own "Mirror Gratitude Project." Once a week, tell someone what you admire about them - clearly, directly, and without strings.
- Reflect appreciation where it's rarely given - to quiet contributors, behind-the-scenes teammates, or even strangers.
- Watch how your own brain responds: You'll begin to **notice more** that's worth appreciating.

🧠 Future-Proof Your Brain Insight
Want to elevate your environment? Model what you want to see - appreciation, positivity, support. Mirror neurons will do the rest.

"We mimic what we see. So, let's show each other the good stuff - on purpose. Like the beauty of giving thanks and receiving it."

> *"There are only two ways to live your life. One is as though nothing is a miracle. The other is as though everything is a miracle."*
> - *Albert Einstein*

Theme 9: Technology, Adaptability, and AI

Way 59: Tech Webinars

During the Covid-19 in Canada the entire province of Ontario was locked down for about 2 years. While most of us were used to online meetings, we were never used to exclusively learning online and we most certainly were not used to juggling having our children at home and monitoring their "asynchronous learning" which was the politically correct code for "home schooling." The ability to adapt during the pandemic lockdown and learn complex information through tech webinars was an unfamiliar experience for many people. From elementary school to university students and professors, everyone was forced to adapt lesson plans, testing, and even how to work together as groups. Professors had to learn to use online platforms, virtual whiteboards, and manage online chat groups…. not to mention the chaos of trying to police plagiarism and cheating on exams.

What most people don't realize is that while virtual learning created a pathway to further democratization of knowledge it also meant that the ability of the brain to remember information learned in person was impeded. Did you know that you are ~70% more likely to remember what a person said when you see their face? This is because ancient evolutionary mechanisms still exist in the brain to associate a face with an important memory. Turning off your camera is the absolute worst thing to do if you want people to listen to you or remember your contributions. This scientific fact might be unpopular, but you cannot cheat your hippocampus.

🧠 Insight: Your Brain Learns Better When It Expects to Engage

Research shows that **passive exposure** to information - just attending a webinar or watching a video - simply isn't enough for deep learning. A 2013 study by Fiorella and Mayer found that learning improves dramatically when people expect to teach someone else. This effect is linked to **social presence**: the feeling that someone is relying on you to communicate information triggers higher cognitive effort, deeper processing, and better memory retention.

In other words:

- **Turning off your camera, multitasking, or passively attending webinars weakens learning.**
- **Preparing to explain, discuss,** or **teach** what you've learned to someone else, forces your brain to organize and store information more effectively.

Ancient brain mechanisms still prioritize **face-to-face engagement** for memory formation - even over virtual platforms. Seeing faces and expecting interaction activate mirror neurons and attention networks that otherwise stay dormant during passive consumption.

Try This

- Keep your **camera on** during virtual learning sessions to strengthen your brain's attention circuits and increase your peer's probability of remembering you and your input.
- After every webinar, **explain the main points** out loud - even to yourself - as if you were teaching someone else.
- Prepare **one question** or insight before every online meeting. Anticipating active participation boosts retention.

Future-Proof Your Brain Insight

In the digital world, passive exposure is not enough. To truly learn, your brain needs social presence, anticipation of engagement, and face-to-face cues - even through a screen. *The future of learning isn't just online - it's interactive, participatory, and anchored in how the brain evolved to absorb and transmit knowledge.*

"The real problem is not whether machines think but whether men do."
- B.F. Skinner

Way 60: Learn to CODE

You don't need to be a software engineer to benefit from learning code. You don't even need to build anything. Being able to speak the lexicon of coding and engineering transforms your understanding of it and solving business problems. When you understand how the wiring works – the building makes so much more sense to you. Perfect analogy – finally building that giant Star Wars Lego spaceship for your kids after you've had a few years of Lego architecture under your belt!

Learning to code rewires how you think. In today's world, you can build websites, apps, and automation with drag-and-drop tools. Lovable AI is one such platform that allows for what's termed "vibe coding." As an AI application builder myself, I've always felt it would be wonderful if more people could contribute to designing new IP. But "vibe coding" doesn't teach you anything about security or regulatory compliance protocols. It doesn't help you understand IT infrastructure.

Understanding the logic beneath the surface - the loops, conditions, systems and structures of code and algorithms - changes how you approach problems, even in daily life. The ability to articulate solutions with the lexicon of coding enables you to unpack technological dilemmas and sustainable solutions. Lovable AI is fantastic for creating wireframes and general workflow depictions – you simply use simple language to explain what you want to achieve in an application. Unfortunately, this doesn't mean that it's necessarily feasible to build.

🧠 Insight: Coding Teaches Systems Thinking

Coding languages like Python, Java, or even HTML and CSS train your brain in **pattern recognition, modular thinking**, and **stepwise logic** - all of which engage your prefrontal cortex, the same part of the brain responsible for planning, decision-making, and abstract reasoning.

Research has shown that learning programming activates areas tied to language processing and mathematical reasoning, particularly the **left inferior frontal gyrus and posterior temporal regions** - indicating that code isn't just technical. It's linguistic. It's logical. And it's deeply *cognitive*.

Here's what that means:
- **You don't have to master code - just learn enough to see the wiring.** Even basic understanding of how code flows, strengthens problem-solving capacity.

- **Coding fluency builds metacognition.** You start seeing how systems operate - and how you operate within them.
- **It makes tech less mysterious.** When you know how things are built, you stop being a passive user - and start thinking like a builder.

Try This
- Take a beginner's course in Python or JavaScript - even just a 5-day challenge.
- Play with a simple coding exercise like automating your to-do list or building a calculator.
- Pair it with a visual - use a flowchart or logic tree to map how decisions get made.
- Build your own GPT. GPT just stands for Generative Pre-Trained Model. It's free and you can do this on OpenAI. Learning to configure your own GPT, uploading reference documents and training tools, training it on responses – all of that is hugely helpful in disseminating many of the false narratives and fears around AI destroying humanity and taking all of our jobs.
- Learn about the differences in LLM approaches – mainly GANs and VAEs (generative adversarial networks and variational autoencoders. Even after building AI tools for 20 years, I learn something new everyday!

🧠 Future-Proof Your Brain Insight

The more fluent you are in how systems work - human, digital, or hybrid - the more agile your brain becomes. *"Code is just another language. And every new language you learn opens a new way of thinking."*

Way 61: Human First, Tech Second

GenAI can write your emails and email responses. Schedule your meetings. Summarize your readings. Even sketch your ideas. But here's the truth:

If you're not careful, you'll outsource your thinking before you've even had one original thought. GenAI tools are brilliant - but they're *not* your brain. They're trained on what already exists and they do have circular logic tendencies. Even humans do. Your brain is designed to generate, not just replicate - and creativity, imagination, even boredom are crucial ingredients to maintaining that ability. GenAI is trained on billions of documents and human contextual knowledge – most of it in English. Although tech players like Nvidia are working to diversify language sets and the nuances of syntax in different languages. In 2025 Nvidia announced the Granary Dataset – over 1 million hours of unique data from multiple languages for AI audio training. Granary is an open-source dataset. This means that anyone can access this data – even threat actors or fraudsters.

Discussions of audio datasets invariably lead us to talk about the changes we are seeing in human language as a result of GenAI-interaction. When humans communicate with each other, we use specific types of language and styles of speaking or formulating our thoughts. This line of ideation encourages us to consider whether Human-AI interactions are creating a new type of language and thought, whereby we maximize cognitive circuitry and capacity via automation, thereby replacing that processing churn of repeated data. There are some major concerns around human beings deliberately choosing AI interaction OVER other humans. One main challenge is "intellectual levelling" - where all human beings start to be collated into a few distinct persona sets based on outsourcing all our communication construction to GenAI. Envision a future of AI speaking to AI with human beings farther apart than ever before.

Widespread access to GenAI tools presents a nascent 2-3 years of data to analyze the impact to human language. With GenAI use specifically there are 3 crucial mechanisms at play raising "Human First" concerns: 1) complete lack of user training 2) moving beyond automation to creation and outsourcing thought and 3) human-AI interaction and integration to the extent of humans choosing AI engagement over human engagement. Before GPS we used maps, before the internet we navigated the dewy decimal system – humanity had a scaffolding mechanism in place for "manual" solutions to solving challenges of navigating or research. What exists to create new ideas and thoughts for us before GenAI? The frontal cortex. Once again remember – Human first, Tech Second.

💬 Insight: Creation Before Automation

Research shows that defaulting to automation too early short-circuits deep learning. When your brain doesn't wrestle with ideas - doesn't get messy, make errors, or stretch its frontal lobe - you **weaken cognitive flexibility** over time. AI is best used after you've sparked your own insight. Why? Because your **prefrontal cortex** (the part of your brain responsible for originality, focus, and innovation) only fires when you engage deeply, reflect, or play with ideas that aren't spoon-fed.

Here's what that means:
- **Prompt engineering isn't the same as thinking.** Let AI enhance, not replace, your ideation.
- **Your originality is the asset.** If you're just remixing, you're not creating - you're compiling.
- **Over-reliance dulls your edge.** The easier it gets to skip thinking, the more intentional you need to be about protecting it.

Try This

- Build a "Human First" Tech Policy: Before asking ChatGPT or any tool, ask yourself:
    1. Have I thought through this first?
    2. Is there a point of view I want to try before searching?
    3. Am I using AI to *amplify* or to *avoid* the hard work of thinking?
- Set a delay: Give yourself 10 minutes to brainstorm before using tech tools.
- Revisit your analog side - sketch, voice record, or write on paper before you digitize.

💬 Future-Proof Your Brain Insight

Creativity is a muscle - and muscles don't grow when you outsource the reps. Use AI as your assistant, not your author. Your brain is still the main character. Train your AI tools on your persona, challenge it, validate what it feeds back to you and always remember that you possess MORE active neurons that any GenAI tools ever can – you just have to use them.

> *"The only way to get artificial intelligence to work is to do the computation in a way similar to the human brain."*
> *- Geoffrey Hinton, The "Godfather" of AI*

Way 62: Anti-Google

"Just Google it." Three words that have quietly reshaped how we think, learn, and remember ever since the verb "googling" entered the Merriam-Webster dictionary in 2006. By always defaulting to Google as your only source of validity in any debate, you've definitively outsourced your curiosity. Before search engines, we had to sit with uncertainty patiently. We guessed. We debated. We experimented. We wrestled with memory and logic and fragments of what we'd once read in a book we could not quite name. That mental wrestling? It's where the brain builds *resilience*.

In 2011, researchers from Columbia University decided to delve into analyzing how we process information found online. What they discovered is that despite your best efforts, you are significantly more likely to forget something you "googled" (Sparrow et al, 2011). They referred to this as the "Google Effect." So, while the internet is a fantastic tool for speed of access, it's not necessarily the best medium by which to entrench data in your memory

💭 Insight: Instant Access Undermines Cognitive Effort

Studies on the "Google Effect," (also known as **digital amnesia**) also revealed that not only did it matter if the information you sought was in fact easy to find, but also whether you yourself BELIEVED it was easy to find. That belief was just as likely to impede your memory retention – almost as if your brain was subconsciously signalling: ***"it's not that hard to find this data again, so why store it for long-term reference?"*** In short, the easier it is to look something up, the less effort your brain puts into remembering or even understanding it. This matters because memory, attention, and reasoning are strengthened through struggle. Cognitive effort is the brain's gym - and Googling everything is the equivalent of hiring someone else to lift your weights. You might notice a recurring theme here between GenAI and Googling – it all circles back to cognitive friction and why your brain needs it to thrive. Contrary to many of the "feel good" memes you see on your social media feed – ease, relaxation, and complete absence of stress is not the stuff of a future-proofed brain.

Here's what that means:
- **The brain needs friction.** Instant answers weaken the very muscles we need for creative problem-solving.
- **Critical thinking is a practice.** If you never sit with uncertainty, you'll never get good at navigating it.
- **We're raising a generation of shortcut-seekers.** But your brain grows from *slow thinking*, not just speed scrolling.

Try This
- Go "Anti-Google" for an hour a day. Commit to solving a problem, making a decision, or sparking an idea without searching it.
- Ask a friend instead of Siri. Or better yet - guess, debate, play with wrong answers first.
- Teach your kids this too: When they say "I'll just Google it," ask them to make three guesses first. You'll build stronger neural pathways *together*.

Future-Proof Your Brain Insight

Uncertainty isn't failure - it's where thinking begins. The longer you can tolerate not knowing, the sharper your brain becomes. *"Sometimes the best search engine is your own mind - messy, biased, beautiful, and deeply human."*

> *"To be fully alive, fully human, and completely awake is to be continually thrown out of the nest. To live fully is to be always in no-man's-land, to experience each moment as completely new and fresh. To live is to be willing to die over and over again."*
> *- Pema Chödrön*

Way 63: Device-Free

Your phone isn't inherently evil. Tools are often crucified by the public when human beings abuse them and leverage them for ill-devised plans. A hammer can be a tool or a weapon – it's not the hammer to blame when it's used to harm someone. Your phone, however, seems to be everywhere and is such a seemingly innocuous tool that it's easy to forget the massive danger it can present to the brain.

It's in the kitchen while you look up recipes or set a timer. 45-60 min in the bathroom while your legs go numb as you get lost in Instagram. It's by your bedside – beckoning to you with its eerie glow "just one more doom school." In the car as you use it for navigating somewhere via Apple CarPlay or Android Auto. Even while you're talking to someone you love, there's your phone, a consummate and unapologetic third wheel.

If you don't set boundaries with your devices, your brain never gets a break. As part of a behavioural experiment on myself in 2024 I turned on screen-time tracking and didn't limit myself at all with my phone. In one week, I spent 49 hours across Instagram, Facebook, LinkedIn, Emails, Safari, YouTube, and all the apps you can imagine. Today, I set a 2-hour screen-time limit on my phone for social media apps, and I limit access to my phone after 6pm – both by physically placing my phone far away from my reach and by putting a screen-limit passcode on my device. Of course, I don't always adhere to the rules I set and sometimes I do need to access my phone after 6pm – but I set the intention to stay off my device and take a break.

Screens don't just compete for your attention - they **fracture** it. And constant low-level stimulation - the pings, the scrolls, the passive checking - keeps your nervous system in a mild but persistent state of alert. You may not notice it, but your brain does. You're like a helpless gazelle on the tundra waiting to be pounced on at any moment by likes, reactions, and comments from other people. Not only does it skew your attention, but it also skews your perception of reality and time.

🧠 Insight: Your Brain Needs Downtime to Integrate

Neuroscientists have found that during *mind-wandering* - the unfocused, idle states when you're not consuming or reacting - the brain activates the **default mode network (DMN)**. You'll notice the DMN comes up a lot in these ways to future-proof your brain. This network is essential for reflection, memory consolidation, and creativity.

When you're always looking at a screen, your DMN doesn't get to light up. You're always reacting. Never reflecting.

Here's what that means:
- **You need screen-free space - not just time.** Designating physical zones sends your brain a spatial cue to shift states.
- **Silence isn't wasted time.** It's where insight forms, memories settle, and emotional regulation resets.
- **Your brain craves variety.** Rotating device-free zones (kitchen this week, bedroom next) keeps the boundary fresh and intentional.

Try This
- Create one device-free zone in your home. Start with a place that feels most human - your dinner table, your bed, or your bathroom.
- Rotate the zone every week. Make it a challenge with your partner or family.
- Put a visible reminder in that space - a bowl for phones, a sign, a candle. Something that signals: Here, we think. We connect. We pause.

🧠 Future-Proof Your Brain Insight
Your brain needs white space to do its deepest work. Don't let your devices crowd it out. Try going even one full day without using your phone, at all. It's remarkably revealing as to just how addicted you've become to that tiny screen. *If your phone is in every room, your thoughts won't be.*

"If it keeps up, man will atrophy all his limbs but the push-button finger."
- Frank Lloyd Wright

Theme 10: The Soft Life and Dreaming

Way 64: A Staycation isn't the same as a VACATION

Vacations aren't just about relaxation - they're a full-system neurological reset. Unlike staycations, which often fail to meaningfully disrupt our inner monologue, true travel introduces the brain to a cascade of novel stimuli that recalibrate our cognitive and emotional systems.

From a neuroscience standpoint, travel activates multiple regions of the brain. The **hippocampus**, responsible for memory formation and spatial navigation, is stimulated by new environments and unfamiliar geographies. The **prefrontal cortex**, our center for planning, decision-making, and goal setting, is re-engaged as we navigate unfamiliar transit systems, languages, or customs. Meanwhile, the **mesolimbic dopamine system** - which governs anticipation, pleasure, and motivation - is triggered in the days leading up to travel, as well as during moments of awe, surprise, and discovery. These systems don't simply enjoy novelty - they rely on it to maintain flexibility, creativity, and resilience.

In contrast, during a staycation, your brain is still surrounded by cues tied to routine. It processes the same smells, sights, and ambient noise as it does on a typical workday. As a result, familiar environments lead to **neural habituation** - the brain relegates much of its sensory input to background processing. There's no shift in attentional demand. This is why you can technically be "off work" but still feel mentally foggy or unrefreshed.

But step into a new setting - say, the moment your bare feet hit the hot sand beneath towering palm trees - and something changes. Assuming that isn't your daily landscape, your nervous system responds with heightened alertness and a slight euphoric lift. That's the **reticular activating system** and **amygdala** lighting up, alert to new stimuli, and reorienting your brain's attention toward the present.

Still, it's not just geography that determines the neurological payoff. Duration matters. Research shows that the **ideal vacation length is at least eight days**, providing enough time to lower cortisol levels, allow the default mode network to quiet, and shift from sympathetic nervous system dominance (fight-or-flight) into parasympathetic restoration (rest-and-digest). Unfortunately, those positive effects begin to taper off after about a month, making the case for regular, intentional time away.

I was asked this exact question during my appearance on *CTV The Social* in July 2024. Melissa Grelo, with her signature blend of insight and curiosity, turned to me and asked, **"So Sarah, how often do we actually need a vacation?"** My answer - which likely made a few HR professionals groan, especially those who were battling Return-To-Office mandates groan - was, **"Ideally? Once a month."**

That suggestion isn't grounded in wishful thinking - it's rooted in the **neuroendocrine timeline**. It takes at least 48 hours for the body to begin clearing excess cortisol and resetting to baseline. Which is also why the four-day workweek is so compelling. The traditional two-day weekend is often consumed by logistical catch-up and anticipatory anxiety. By Sunday afternoon, most people's **prefrontal cortex** is already shifting back into task-mode - robbing us of the restorative space we so desperately need.

In short, your brain doesn't just crave rest - it needs change. And in an age defined by digital fatigue and chronic stress, the most effective cognitive intervention might just be a plane ticket.

🧠 **Future-Proof Insight:**
"Relaxation is local. Renewal is geographic." To future-proof your brain, change your setting - not just your schedule.

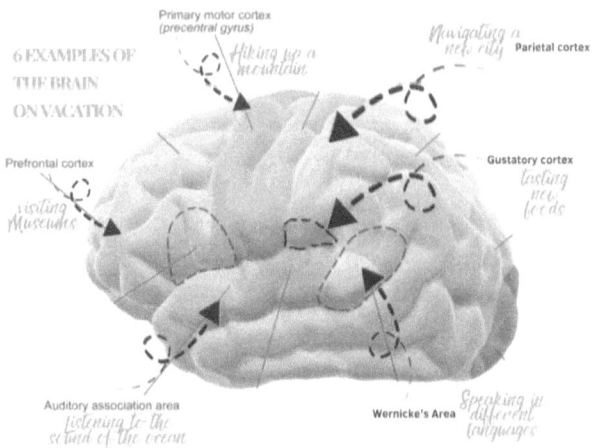

Image: Neurological Impact of a Vacation on Your Brain – Created for CTV The Social, July 2024

Way 65: Paraskevidekatriaphobia – that's a mouthful!

Neuroscience vs. the Friday the 13th Effect

In September 2024 I joined NBC News in South Florida to unpack this long long long word. Paraskevi means Friday, Dekatria means thirteen and Phobia means fear. It translates into a fear of Friday the 13th specifically. The term was coined by psychotherapist Donald Dossey and he joked that if you could pronounce the word you'd be instantly cured of the affliction!

As I walked into those NBC Studios in South Florida I too had superstitions about my first time on live news in Florida – I wore a new outfit, I kept it a secret right up until that morning because I didn't want to "jinx" it. The segment happened right around a big football game for the Miami Dolphins and the point of unpacking the science behind superstition was to understand why exactly it's so very easy for humans to adopt these belief systems, especially around big sports games.

Superstition is just cognitive noise. The brain's pattern-seeking bias makes us cling to false predictors. But superstition also activates the amygdala, keeping us in low-grade fear loops. Long-term, this reinforces neural rigidity - and blocks critical thinking. On every occasion that the superstition is "proven" the brain stores a memory, sometimes called "confirmation bias." This also happens when we have a dream about something and then it happens – our primal reaction is to start imagining that that have secret psychic powers. That possibility is supremely more interesting than unpacking the statistical probability of all the times we had dreams about something that didn't happen, or something happened that we didn't dream about.

We also experience something called attribution bias – the brain trying to make sense of things. We see a black cat in the morning and then have a bad day – and inaccurately we highlight that black cats must be bad luck, when in reality there are very likely far more days when we received bad news and didn't see a black cat! It's all very unhinged logic when you say it aloud.

In examining superstition, it doesn't show up in neuroscience or evolutionary biology research as a prevalent theme. It does however make appearances in psychology – most notably by behavioural psychologist BF Skinner, first in the 1940s when he examined potential superstitious behaviours in pigeons and their response to food. More recent research from 2008 unpacks the purpose of superstitions and suggests that superstition is an inevitability in all living things – an outcome that results from adaptation and our need to attribute purpose and reason, even if our explanations are completely lacking in logic.

🧠 Future-Proof Insight:

Every time you break a superstition, your brain gets braver. It stops seeking the lazy explanation and begins looking for facts and data to support what occurred. But your brain needs to be comfortable with expending that extra effort, and it needs to be trained to see the value in additional cognitive load. *To future-proof your brain, challenge inherited and generational myths - your logic deserves the last word.*

"I am well-educated enough not to be superstitious, but I am superstitious." - Fyodor Dostoyevsky, "Notes from Underground," translated from the Russian by Constance Garnett, 1918

Way 66: 8 hours a day keeps the…

You deserve a break. But your brain doesn't want to completely power down. I for one have never been good at taking naps, not at 4 months old and not at 40 years old. The caveat with rest is that total shutdown isn't the goal for everyone. Gentle stimulation is. Rest is essential - it lowers cortisol, restores energy, and supports emotional balance. But when rest turns into mental emptiness, especially for days at a time, your prefrontal cortex - the part of your brain responsible for decision-making, focus, and innovation - starts to lag. Anytime we go through an extended period of low neural stimulation, the brain begins to "spring clean" through a process called neural pruning. In the case of your brain the adage "use it or lose it," is accurate.

The common assumption is that everyone needs 8 hours of sleep. This came from the 5-day work week and structures around social mechanisms of scheduling. Perhaps you've met people who don't feel content with 8 hours, they thrive on 3 hours – and you wonder what exactly is wrong with them. Possibly they are sleep deprived and running on pure adrenaline, but it also might be explained by genetic factors. Some people possess a rare genetic mutation of what's called the DEC2 gene, which means that they really only need a precious few hours of rest and they're rejuvenated.

The DEC2 gene was first discovered in 2009 during a study of a mother and her child. It was the first study to reveal that this tendency to need less sleep could be genetic (He et al, 2009). It's a fantastic example of why proclamations of "the perfect diet" or the "perfect lifestyle" are to be cautiously believed. For carriers of this genetic mutation there is a neuropeptide produced in the hypothalamus called orexin. Your hypothalamus controls your body temperature and even hunger – you can envision it as a control center for keeping the body stable. Throughout the day orexin naturally fluctuates in your system, as it decreases you become sleepier. People with narcolepsy, who randomly fall asleep, they have an autoimmune reaction to orexin and those cells are destroyed. Sleep is not one size fits all. And similarly, neither is how much stimulation a person needs to feel cognitively optimized.

- Rest isn't the absence of thought. And it's not necessarily always sleep. It's the shift from *performance* to *play*.

- If taking a nap sounds like torment remember that "doing nothing" is different than "doing something light." And the latter is better for your long-term brain health.

Try This
- Give yourself a sleep holiday – a week where you don't have to match your sleep patterns to anyone else but you. Or if you're having a more significant challenge with insomnia or sleep, it might be time to visit a sleep clinic.
- Think "restorative stimulation," not just silence. If you find yourself resisting sleep and silence – focus on moving away from patterns of always "being productive"
- Stop trying to match your sleep patterns to what you've been told is "normal" – find your own level of rest that works for you.

Future-Proof Your Brain Insight

On your next vacation, give yourself the space from alarms, schedules, and routines to discover your natural sleep pattern. There are genetic tests for the DEC2 mutation, but you can also just observe how much sleep you need when you're away from structured life. You just mind find that 8 hours is too much for you.

> *"We are such stuff as dreams are made on; and our little life is rounded with a sleep."*
> *- William Shakespeare*

Way 67: SLEEP...Quality over Quantity

While only 1-3% of the population has the DEC2 gene mutation discussed in Way 66, we place too much importance on counting hours of sleep. Morning routines get all the attention - but evening neural deceleration may matter even more. The quality of your sleep isn't determined the moment you close your eyes - it begins hours earlier, shaped by how you transition into sleep. Proper evening routines strengthen slow-wave sleep (deep sleep stages), activate the glymphatic system to clear neurotoxins from the brain, and stabilize your emotional regulation by calming the reticular activating system (RAS).

For years, after my divorce, I lived in chronic sleep deprivation. As a lone parent raising a toddler, I often slept just three to four hours a night. I would work all day, parent in the evening, run my household, then dive back into work once my son was asleep - collapsing from exhaustion well past midnight. There was no deliberate deceleration, no "winding down." I believed that passing out from sheer fatigue meant I was getting "good enough" sleep. But I was wrong. It turns out that I do have the DEC2 gene but even for someone like me, 3-4 hours of low-quality sleep is not enough.

Exhaustion sleep is not restorative sleep. When you force your body to shut down under stress, you disrupt the natural processes your brain needs to heal overnight - especially the phases of slow-wave sleep that are critical for: *clearing toxins, consolidating memory, and regulating mood.*

💬 Insight: Sleep Is Brain Maintenance, Not Just Rest

Recent research by Dall 'Orso and colleagues (2021) highlights the crucial role of the **glymphatic system** - a specialized clearance system that activates primarily during deep sleep to flush out neurotoxins, including metabolic waste products linked to Alzheimer's and other neurodegenerative diseases. The study found that poor slow-wave sleep impairs this brain-cleaning system, increasing the buildup of harmful substances that compromise cognitive function over time.

In simple terms: If you don't decelerate properly before bed - if you just crash from overstimulation or stress - your brain's natural housekeeping functions can't do their job.

Try This

- **Dim your lights** two hours before bed to signal your RAS (reticular activating system) that it's time to wind down.
- **Stop doomscrolling** and avoid screens 60–90 minutes before sleep - blue light delays melatonin release and overstimulates your brain.
- **Use breathwork or body scanning** for 5–10 minutes to shift your nervous system into parasympathetic (rest-and-digest) mode.
- **Create a ritualized descent**: repeat a short series of calming actions each night so your brain starts associating those behaviours with safe sleep entry.

Future-Proof Your Brain Insight

To future-proof your brain, stop treating night as a leftovers zone. Design your descent. Good sleep isn't about collapsing from exhaustion - it's about **actively designing** your path into slow-wave recovery, every night. *"The quality of your sleep begins hours before you close your eyes."*

Way 68: As the song goes… "Must be Love on The Brain" – Rihanna

We say "I love you" to a partner, to a parent, to a pet - even to a slice of cake. But your brain? It keeps the receipts. Different kinds of love light up your brain in diverse ways. In English, we only have one word for love. But in Ancient Greek or Latin, there were many: *eros* for romance, *storge* for family love, *philia* for friendship. And neuroscience backs this up - your brain knows the difference.

There is also a social mechanism at play where we have been increasingly shamed for comparing distinct types of love or trying to quantify love. Can you love a love and a child in the same way? Your partner or a stranger? If we're strictly speaking in neuroscience terms – your brain does have a spectrum of intensity for types of love. And in current times there is a new kind of love being articulated, people falling in love with AI personas – a kind of love that for many people seems unimaginable and for others is classified as "the most pervasive love they've ever felt." With GenAI becoming increasingly skilled at understanding the nuances of human emotion it's not shocking that people are falling in love with AI personas that seem to understand their needs, wants, and quirks better than any real human being.

💭 Insight: Love Lives in Multiple Neural Neighborhoods

Researchers at Aalto University in Finland used MRI scans to observe how the brain responds to six types of love: romantic, parental, platonic (friendship), love for strangers, love for pets, and love for nature. The results? While similar regions were activated across all types - including the precuneus, temporoparietal junction, and basal ganglia - the intensity varied drastically.

Here's what they found:
- Love for children activated the brain most intensely - especially the striatum, your deep reward center. Which didn't even show up for romantic love.
- Love for strangers and nature? These were the least intense - lighting up mainly your visual and reward centers.
- Love for pets boosted activity in social-processing regions - suggesting that pet ownership enhances your brain's empathy circuits, even if it doesn't hit as hard as human love.

Each kind of love has a signature neural footprint, tapping into different blends of oxytocin, dopamine, and vasopressin - the brain's trust, reward, and bonding hormones.

The Love Pyramid:
- Romantic love is dopamine-rich - thrilling, motivating, risk-taking.
- Parental love is oxytocin-fueled - grounding, selfless, biologically intense.
- Friendship and social love activate empathy circuits - emotionally regulating and stabilizing.

Try This
- Build a "love map" - reflect on the people, animals, and experiences in your life that activate *different kinds* of love.
- Don't just chase romantic highs. Seek out the love that stabilizes you, the one that challenges you, the kind that makes you laugh without needing anything in return.
- Feed your brain a balanced love diet - it needs the full range to thrive.

💬 Future-Proof Your Brain Insight

Love isn't one-size-fits-all - and your brain knows it. To future-proof your neural networks, nurture every kind of love: parental, platonic, romantic, and even the kind that comes on four legs. *"Different loves light up different parts of you. Give your brain the full spectrum - and it will love you back."*

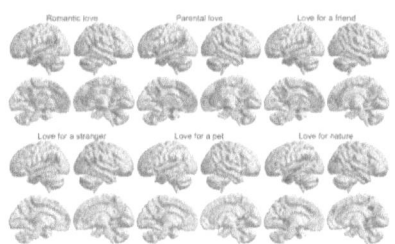

Image: Different Types of Love Light up different regions of the brain. (Rinne, 2022)

Way 69: Make Uncertainty Your Training Ground

Your brain hates not knowing – in Way 29 we first touched on this. In fact, the brain would rather settle for a *bad* outcome than sit in the discomfort of "maybe." When you read that out loud, it sounds ridiculous, but it's unfortunately true. Avoiding ambiguity makes your brain rigid. Embracing it makes it resilient. Studies show that people will choose a known negative over an uncertain outcome, simply to reduce anxiety. That's how deeply wired our brains are for closure. But here's the problem: life doesn't work that way. Neither does growth.

In my childhood there was a lot of ambiguity. From age 3 onwards I didn't know if my parents would have an argument about shared custody, weekend visits, holidays, or who would show up to school events. As an adult, I abhor uncertainty – I dislike amorphous discussions that lead nowhere, and I avoid meetings just for the sake of meeting. I thrive in knowing the objective, the desired outcome and the path forwards. My reaction to uncertainty is a neural trigger that is predicated upon feeling a lack of safety and stability. It means that for people like me, we must work that much harder on being at ease with the unknown.

Insight: Ambiguity Is a Workout - Not a Threat

Research published in *Frontiers in Psychology* (Hope, 2021) found that navigating uncertainty strengthens key cognitive networks - especially the anterior cingulate cortex (linked to conflict monitoring and emotional regulation) and the prefrontal cortex (your decision-making and planning hub).

When you sit with ambiguity - when you don't rush for resolution - your brain becomes more flexible, more focused, and more emotionally steady. I for one, do not enjoy the ambiguous, the nebulous, the uncertain or ungrounded – deliberately sitting in mental space that is unclear and unstructured feels like torture. But when I finally overcome those instinctual feels of abhorrence, my brain starts evolving.

Here's what that means:
- Uncertainty tolerance is linked to stronger mental health. People who can pause before reacting - and sit with ambiguity - are more adaptive and less anxious.

- Ambiguity is neuroplasticity in action. Each time you delay the need for a fast answer, you reinforce myelin sheaths - the protein-rich conductors that help neurons fire more efficiently.
- It's like resistance training for your mind. Ambiguity is your cognitive dumbbell. Pick it up often.

Try This
- Practice "wait time." When facing a decision, pause before rushing to act. Even 30 seconds of reflection can shift neural patterns.
- Gamify ambiguity: Ask "What if I sat with this uncertainty for a day? A week?" Treat the unknown like an experiment, not a danger zone.
- Expose yourself to small daily discomforts - from new routines to open-ended tasks - to build your ambiguity threshold.

🧠 Future-Proof Your Brain Insight

The ability to hold uncertainty without panic is one of the strongest predictors of resilience. Your brain doesn't need perfect answers. It needs practice staying present without them. *If you can sit with uncertainty, you can survive anything.*

> **Uncertainty is the only certainty there is and knowing how to live with insecurity is the only security."**
> *– John Allen Paulos*

Way 70: Sunny with a Chance of Rain, Clouds, and Thunder?

You check the weather every morning - but do you actually *feel* it? You say, "hey Google, what's the weather today," thinking that the only impact is going to be whether you wear open-toed shoes or closed-toe. Or if you're Canadian, for 9 months you ask "will it rain, be overcast or snow today? (little bonus Canadian humor for you there)"

The weather isn't just outside you. It's inside you. Your brain doesn't just passively observe sunlight, storms, or seasonal shifts. It responds. It recalibrates hormones. It resets timing systems. But only if you let it. In Way 39-42 you learned about seasonal alignment, and why New Years Resolutions are least likely to succeeds in January. The weather doesn't always coincide with the seasons – and as human beings shift towards different modes of work people are more often exposed to unpredictable or unfamiliar weather.

Typically, we praise sunlight exposure, but people react to storms and solar events as well – and not only because of the mystical and exciting stories surrounding them. Storms impact barometric pressure and for people who are particularly sensitive, they'll notice that during thunderstorms they tend to have headaches or migraines. When air pressure drops suddenly, there is a dissonance in your sinuses and that can cause intense pain. Think about what happens when you're on a plane and your ears start popping as the plane ascends or descends. Barometric pressure also impacts cognitive function – sometimes negatively or positively.

🧠 Insight: Nature's Rhythms Are Your Neural Rhythms

We often think of ourselves as separate from the environment - like the weather is something we either avoid or endure. But your brain evolved with the sky. A 2016 review in *Frontiers in Psychology* (Lambert, 2016) found that sunlight exposure directly influences serotonin production, melatonin release, and cortisol rhythms - all of which regulate mood, sleep, and energy.

Here's what that means:
- Sunlight exposure increases serotonin, your mood stabilizer. This is especially critical in winter months or regions with fewer daylight hours.

- Morning light suppresses melatonin and raises cortisol at the right time - helping to lock in your circadian rhythm and enhance alertness and focus.
- Nature is neurological. Over-sheltering yourself from light, rain, wind, or snow weakens the body's circadian feedback loop - leaving you foggy, fatigued, or flat. Sun all the time is not the answer either.

Try This
- Spend 10–15 minutes outside within an hour of waking. Even on a cloudy day, natural light is stronger than most indoor lighting.
- Don't hide from the weather. If it rains, let it. Walk without an umbrella occasionally. Let your skin feel it. Let your brain register it. As they say in Germany: "You are not made of sugar! You will not melt."
- In winter, prioritize natural morning light exposure or consider full-spectrum light therapy if you're in a low-sunlight area.
- Buy Seasonal Affective Disorder (SAD) lamps to help counteract the effects of SAD. *Refer to Way 41 for a deeper discussion of SAD

Future-Proof Your Brain Insight

Mood, focus, sleep - they're not just mental. They're *environmental*. Sync your biology with the sky, and you'll feel the lift - naturally, rhythmically, and neurologically. *"Your brain is a solar-powered, rain-responsive, seasonally adaptive system. Let it live accordingly."*

Way 71: Learn how to do the Things You Think You're Bad At!

"I'm just not a numbers person." "I can't draw to save my life." "I've never been good at languages." "Gardening is really not my thing..." Sound familiar?

The things you've avoided might be the very things your brain needs. Most adults hit a comfort zone and settle there. We stick to what we're already good at - and avoid the skills that once embarrassed us. But when you push into the "I'm-bad-at-this" zone, your brain lights up in ways that easy tasks never could. You're cracking open the neural activity jackpot!

For years I avoided gardening. The worms. The dirt. The waiting for things to grow. It was viscerally unappealing to every fiber of my being. My safe space was high heels, pencil skirts, boardrooms, keynote stages, or being in "mom mode." I told myself, **"You grew a human, no need to grow plants!"**

Then one day, I looked out at my backyard and took a trip to Sheridan Nurseries in Toronto – loaded up my car with tulips, hyacinths, lily of the valley, and daffodils, and decided to make something beautiful. A few hours later? I was covered in mud but loving the experience of making a flower bed grid, designing entrance way urns, and looking forward to seeing more flowers sprout in a few weeks.

Insight: Neuroplasticity Loves Discomfort

Struggling through something new - especially something you've labeled "not for you" - activates the **hippocampus** (your learning and memory hub) and promotes **new grey matter growth**, particularly in the **prefrontal cortex**, which governs planning, decision-making, and identity.

This isn't just skill acquisition. It's **self-concept rewiring**. You're not just learning how to dance or do math - you're teaching your brain a new story about who you are.

And when you bring humor and humility into the mix - when you laugh at yourself, stay curious, and keep going - the learning sticks. Literally. Dopamine spikes, attention improves, and emotional resilience strengthens.

Here's what that means:
- **Avoidance keeps you cognitively static.** The "I'm bad at that" script is a mental cul-de-sac.

- **Trying rewires your story.** Even small wins retrain how your brain sees your abilities.
- **Discomfort ≠ danger.** In fact, it's the opposite. It's a sign of new neural pathways being formed.

Try This
- Pick one thing you've always told yourself you're bad at - dancing, drawing, public speaking, mental math - and practice it *badly*.
- Track your reaction. Where do you flinch? Where do you self-criticize? That's your neuroplasticity invitation.
- Do it with someone else. Humor and shared failure lower the stakes and increase learning.

💬 Future-Proof Your Brain Insight

The skill you've avoided might be the one that keeps your brain young. Fail on purpose. That's where new wiring begins. Children must be encouraged to retain their ability to be generalists before we ever push them into becoming specialists. *"Discomfort is the price of neuroplasticity. Pay it gladly."*

"Those who have a 'why' to live, can bear with almost any 'how.'" - Viktor Frankl

Way 72: Build a Relationship with Friction

We often treat "easy" as a synonym for "healthy" - especially when it comes to relationships. No tension? No disagreement? Must be perfect, right? A relationship without friction may be comfortable - but it won't stretch your brain. The theme of cognitive friction has arisen several times in this book – but what about visceral human friction? As children we are told to make friends and get along. Social collaboration is good for the tribe. Invite every single kid to the birthday party, make sure no one feels left out. And that makes sense for rudimentary learning pods, but for adults always agreeing, never being challenged or challenging others – that is a cognitive quagmire of mediocrity. Complex human relationships that stretch your cognitive boundaries are going to present you with perspectives that you may dislike, if not outright abhor. This is why it is so very important to train your GenAI tools to disagree with you – so that you don't become addicted to interacting with algorithms that always tell you "you're right."

💬 Insight: Safe Conflict Builds Cognitive Flexibility

Neuroscience shows that healthy disagreement activates your social brain - particularly the **temporo-parietal junction** (TPJ), which helps you imagine other people's perspectives, and the **dorsal anterior cingulate cortex**, which helps regulate emotional conflict and maintain empathy under stress. Relational neuroscience also highlights the power of *presence* - showing up authentically, even in moments of discomfort. According to Dr. Heller and colleagues, the capacity for "attuned presence" - being emotionally available and non-reactive during conflict - **stimulates limbic resonance**, strengthens **attachment circuits**, and deepens emotional regulation over time.

Here's what that means:
- **Easy doesn't equal growth.** When we're only surrounded by people who validate our worldview, our **neural flexibility stalls**.
- **Conflict isn't always a threat.** Safe disagreement - the kind rooted in trust, not attack - strengthens empathy and rewires stress responses.

- **Presence is more powerful than agreement.** Being seen, heard, and disagreed with (gently) sharpens your ability to regulate, reflect, and relate.

Try This
- Reflect on a relationship where you feel safe but are often challenged. That's your **"neuro-gym"** - where friction leads to growth.
- Practice "listening to understand, not to respond." Watch how your brain calms when you release the need to be right, or the need to have an answer already fabricated.
- Invite thoughtful tension: Ask someone you respect to challenge one of your assumptions - and stay open through the discomfort.

Future-Proof Your Brain Insight

A brain that never disagrees is a brain that never expands. The best relationships don't just validate your views - they *reshape* them. *"Friction isn't failure - it's how connection takes shape in three dimensions."*

> *"Honest disagreement is often a good sign of progress."*
> *- Mahatma Gandhi*

Way 73: The True Cost of Perfectionism

Masking - suppressing natural behaviour to appear acceptable - increases **cortisol**, burns cognitive resources, and leads to emotional disconnection. The reward? Social approval. The cost? chronic neurochemical imbalance and burnout. Research shows that tends to be especially common in neurodivergent populations.

This behaviour is also present when people constantly "code switch," adjusting their tone, cadence, and mannerisms to try and "fit in." No one is a perfect fit for every group – and if you expend so much energy trying to be accepted by others there's not much neural energy left to be you. True, you might get a short-term dopamine boost from social acceptance, but the feelings of cognitive dissonance and misalignment will only surface later.

"Be authentic," sounds like cliché advice – but it's sound advice. Stop trying to be perceived as perfect or liked by everyone. The only way to be liked by everyone is to stand for nothing, and compromise on everything – and even your brain knows that isn't good for you. The brain only seeks approval because it ties back to ancient evolutionary mechanisms of belonging, social cohesion and safety. In current times, it's questionable whether being determined to hold true to your views is welcomed all over the world – it isn't – yet for the brain, being challenged only yields neurological strength if you are able to debate your viewpoints without yielding to aggression.

🧠 Future-Proof Insight:
"If you spend all your brainpower blending in, you'll forget who you are." To future-proof your brain, drop the performance. The real upgrade is in radical honesty. With yourself, if not always with everyone around you.

> *"It's in our biology to trust what we see with our eyes. This makes living in a carefully edited, overproduced and photoshopped world very dangerous."*
> - *Brene Brown*

Way 74: Protect Your "Novelty" Drive

Novelty activates the **substantia nigra/ventral tegmental area (SN/VTA),** flooding the brain with dopamine and BDNF. Repetition dulls this system. Even micro-novelty - rearranging a room, exploring a new walking route, trying new spices - strengthens curiosity circuits.

Micro-novelty doesn't have to mean expensive trips to the Aegean Sea and shopping expeditions – even trying new foods and activities can stimulate the ventral tegmental area. For years I avoided the "adult-colouring trend" – I thought the name itself sounded a bit bizarre, but recently I grabbed some pencil crayons and dived into colouring. This experience was totally different than painting or sketching and it ended up being calming. Will I do it again? Likely no. But the novelty of the experience certainly made an impression upon me, and it also helped me to be more creative in my work approaches and solutions to problems.

A predominant reason that novelty is so difficult is that it requires MORE brain power. And the brain does strive for efficiency as much as possible unless its owner pushes it beyond those simple operating mechanisms. Novelty does spike dopamine in the brain however, which in turn leads to higher activity in the frontal cortex and the hippocampus – this is why when we discover a new experience it feels so exhilarating. Navigating the new drives learning and brain "arousal" (Weierich, 2009) – it's as simple as "trying new things" and continuing to do so even when the outcomes aren't always pleasant.

The idea of "the new" is one of the main reasons that there is such a strong resistance to learning to use AI. Many researchers postulate that it is due to job security concerns and the proliferation of "anti-AI" sentiment online, but most new technologies struggle through the stages of adoption because people are inclined to keep doing what works. Investing energy into learning to use a new email, new website, new backend system....new processes, it's all perceived as painful. In the case of AI the pace of change has outstripped anything humanity has experienced previously.

Generally speaking, Industrial Revolutions are considered by historians to occur every 70-100 years. Here's a quick history overview:

1st Industrial Revolution: (c. 1760–1840): Steam power, water power, and mechanization.

2nd Industrial Revolution: (c. 1870–1914): Electricity, steel, and mass production.

3rd Industrial Revolution: (c. 1969–2000s): Computers, electronics, and early automation.

Today many believe, me included, that we are in the 6th Industrial Revolution, which is characterized by Cyber-physical systems, IoT, AI, and smart factories and perhaps most importantly, quantum computing. But within the past 100 years the momentum of innovation has increased. In the 1960s it started to accelerate between automated chess players beating Grand Masters, all the way to robotic vacuums in the early 2000s. With GenAI, the inflection points now occur not within decades, but within months. OpenAI was released in 2022, and today ChatGPT goes through algorithm updates every few months. The human brain was originally built for navigating unknown terrains, over the years of domestication we've lost much of our skills to adjust to rapid change. We've lost much of our neurological grit. The pace of change from 2022-2026 demands that human beings upgrade their neurological operating systems – before the momentum jumpstarts to an even faster pace in the near future.

🧠 **Future-Proof Insight:**
"New doesn't mean drastic. It just means not the same as yesterday." To future-proof your brain, chase tiny surprises - your neurons light up every time. Whether it is learning to navigate AI tools or just planting a beautiful flower and covering your hands that are used to dancing on a keyboard…. with soil instead.

> ***"The pleasure of novelty is by its very nature more subject than any other to the laws of diminishing returns."***
> *- **C.S. Lewis***

Way 75: Buy Yourself Flowers…Like Miley Cyrus Says

It turns out Miley was right: Buying yourself flowers isn't indulgent - it's neurological self-care. Whether someone hands you a bouquet, you're picking wildflowers in a field, or you grab a pre-fab bunch at the grocery store for yourself, having fresh flowers in your home literally boosts your brain chemistry. This isn't metaphor. It's dopamine. Ever since I left home at 18, I've always invested in having fresh flowers in my home. It's about more than just beauty; it's about the easy neurological boost they give you.

Here's how it works:
- Flowers trigger **visual novelty and aesthetic pleasure**, which activate the **mesolimbic dopamine system** - your brain's reward circuit.
- Exposure to natural colour palettes, particularly greens, purples, and yellows, enhances **prefrontal activation** and can reduce **stress reactivity in the amygdala**.
- The **olfactory system** (your sense of smell) connects directly to the **limbic system**, the emotional center of the brain - meaning the scent of flowers can directly influence mood and memory.
- Studies show that patients recovering from surgery need **less pain medication** when exposed to fresh flowers and plants in their rooms. Their **parasympathetic nervous system** (calm state) activates faster.

In a world where so much feels digital, abstract, and transactional, flowers are physical, alive, and temporary - which is exactly why they feel so emotionally grounding. They remind the brain: this moment matters.

🧠 Future-Proof Your Brain Insight
Beauty isn't extra. It's essential. Your brain needs micro-moments of joy to keep the dopamine system healthy and resilient - not just during big events, but in everyday life. *"I don't need a reason to bring joy into my space - I just need to remember it helps me grow."* To future-proof your brain, buy the flowers. Not because you "earned them." Because your brain will **bloom** with you.

"Meditative Rose"
- By Painter Salvador Dali

Way 76: Step Inside a Smarter Space

From the arches of Gothic cathedrals to the clean lines of postmodern towers, architecture doesn't just reflect civilization - it shapes your cognition. Touring architectural spaces engages the **parietal cortex** (spatial navigation), **visual cortex**, and **limbic system**. Geometric shapes like circles and spirals increase aesthetic pleasure and calm, while vertical lines and vaulted ceilings activate **awe circuits** in the brain.

As a child I imagined going to the Eiffel Tower and the awe I would feel – I had seen the beautiful tower in so many images and movies. But when I finally made it to France for the first time at aged 32 with my 7-year-old, I was kind of underwhelmed. What did inspire me? Montmartre, Versailles, the juxtaposition of so many different types of architecture and shapes that were unfamiliar to me. Perhaps most surprisingly stimulating was learning to navigate the underground spiderweb known as the Paris Subway system!

Recently, I took an architecture tour in Chicago and visited the Chicago Cultural Center where the stained-glass dome took my breath away. When we visited Macy's and saw the largest Tiffany-glass ceiling I was speechless. Seeking out these spaces of unique beauty challenges the monotony and familiarity of everyday workspaces. AI can generate fantastic images and videos, but it still struggles with creating the unpredictable. Virtual Reality Headsets have come remarkably far in the past few years, but they still cannot seem to replicate for the human mind what the real physical world provides. That is not an invitation to eschew technology and fun VR experiences, but rather to acknowledge the importance of your cognitive relationship with the 3D Cartesian dimension that your corporeal body occupies.

🍃 Future-Proof Insight:

"Your brain responds to space like it responds to music - some rooms make you think bigger." To future-proof your brain, walk through new spaces intentionally. Let the structure reshape your perspective.

Original Photo – Taken March 2025.
The Tiffany Dome, on the south side of Chicago, is the world's largest Tiffany glass dome, restored to its original splendor in 2008. The dome on the north side of the building is a 40-foot-diameter dome with an intricate Renaissance pattern, designed by Healy & Millet.

Way 77: Choose Luxury Over Scarcity

When you think "luxury," what comes to mind? Labels? Yachts? Five-star hotels? Luxury isn't about status. A frequent misconception of the word luxury is that it must always be equated with being more expensive. Materialism gets a bad rap, particularly in the current system that most of the world utilizes – fiat money. John Locke for example proposed that we avoid fiat money – which is a system in which currency itself has often arbitrarily assigned values. Locke proposed instead that "sound money" was a preferable mechanism of monetary value where money based on the intrinsic value of precious metals.

Your brain has a relationship with currency, whether it is paper, gold, or even seashells as are still used in some cultures. Money represents value and the ability to obtain resources. Your brain pursues resource accumulation – which is often classified as being "money hungry." It's worthwhile to note that pursuit of resources has always been integral to human survival – whether it was the accumulation of land, food, or power.

The relationship we have with money is about signaling safety - to your brain. The human brain isn't wired to plan for the long-term when it feels threatened. Scarcity, whether financial, emotional, or time-based, triggers the amygdala, your brain's fear center. And when the amygdala is running the show, your prefrontal cortex - the part responsible for long-term thinking, future planning, and calm decision-making - goes offline. In an era when many people are panicking over job security during the age of AI, it is difficult for many employees to 1) plan ahead and 2) think about upskilling within their job functions.

🧠 Insight: Safety First, Strategy Second

The only way your brain can truly plan is if it feels safe in the present. That's where the parasympathetic nervous system comes in - the "rest and restore" branch of your nervous system. Practicing moments of intentional luxury - good sheets, soft lighting, unhurried meals, real rest - helps shift your body out of stress mode and back into creative, strategic, expansive thinking.

And the best part? It doesn't take a trust fund. It takes intention.

Here's what that means:

- Scarcity shrinks your vision. You can't dream big when your brain thinks the world is burning.

- Micro-luxuries recalibrate your nervous system. Whether it's a hot bath, a great espresso, or an extra 20 minutes of sleep, these moments signal that you are safe enough to relax.
- Relaxation unlocks innovation. Once your nervous system feels calm, the prefrontal cortex lights up - and suddenly the future feels possible again.

Try This
- Redefine luxury. Think *sensorial safety*: your favorite tea, natural fabrics, quiet, music, touch.
- Build rituals of abundance into your week. A morning without rushing. A midday walk with no phone. A single moment that whispers, "you're allowed to be here."
- When you feel stuck in fight-or-flight, ask: What would signal *safety* to my body right now?

Future-Proof Your Brain Insight

You can't plan while stuck in survival mode. Create space for micro-luxuries - they don't just soothe you; they rebuild the neural pathways that allow you to dream, decide, and design what comes next. And if you need ideas for micro-luxuries – yes you can ask AI for ideas. *Abundance isn't excess. It's the permission to believe in tomorrow.*

> *"The saddest thing I can imagine is to get used to luxury."*
> - *Charlie Chaplin*

Way 78: Understanding The Aging Brain – By Gender

One unavoidable way to navigate our brain being prepared for the future is to accept and incorporate the inevitability of aging into how we structure our plans for brain training. People are living longer and working longer. We are all going to age and as we do it's helpful to understand how hormonal changes are going to interact with changes in neurotransmitters and the resulting behavioural differences.

Men and women do age differently – especially neurologically. Men experience gradual testosterone decline, often leading to slower changes in memory and mood. Women experience a sharper estrogen drop during menopause, which can increase risks for Alzheimer's, mood changes, and cognitive fog. However, women's brains show stronger cross-hemisphere connectivity, which supports adaptability post-menopause. This contradicts a common conception around menopause and brain fog – which is not a symptom for every woman.

There is also a marked sensitivity around discussing age and particularly around any potential negative connotations. One erroneous assumption is that the older a person is, the harder it is for them to learn new tasks or to use technology. However, the single greatest predictor of the ability to upskill on tech is personality, not age. This shows up in brain scans where neurological age and biological age are not always correlated. In fact, they are rarely identical.

This is why conversations around intergenerational workforces are more crucial than ever. Members of the workforce who are older are frequently becoming victim of these ageist assumptions. Worth noting is that GenZ and GenAlpha employees are also victims of reverse ageism – where their youth is immediately correlated with being less experienced, less intelligent, and less capable.

In considering ageism around cognition there is also research that shows men and women are demonstrating vast differences in how they explore AI in their work. In a 2024 Forbes study approximately 10,000 "desk workers" were surveyed. The survey found that Gen Z men are 25% more likely to have tried AI tools compared to Gen Z women. Overall, more men are trying AI than women, the survey found, with 35% of male respondents saying they had tried AI for work, compared with 29% of female respondents.

"This is something that we should be keeping an eye on," said Christina Janzer, head of Slack's Workforce Lab. "My hypothesis is that the people who are using it today are the people who are going to help shape the future of it. We want those people to be representative of our entire population. That's not what we're seeing today. This is a big opportunity for leaders to understand that and to course correct."

Beyond age and gender, the report did find that workers of colour were using AI more than white workers, with 43% of Hispanic, 42% of Black and 36% of Asian American respondents reporting they'd tried AI tools at work, compared with 29% of those who were white. The important caveat in analyzing this study is not to criticize any group - but rather to equalize informed access to AI, to give access to learning tools and to ensure that this disparity is eradicated as early as possible in the AI Adoption Trajectory.

🧠 **Future-Proof Insight:**

Aging isn't decline - it's recalibration. But you need to know the code in order to future-proof your brain's roadmap for aging. *To future-proof your brain, tailor your self-care by sex and stage - your neurobiology deserves nuance.*

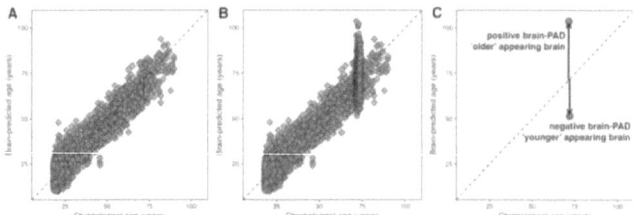

Image: Chronological age and neurological age are not always correlated (Cole, 2019)

Way 79: Ditch The Attention Economy

The moment you wake up, your hand reaches for your phone. You're not alone. Most people do it - often before even getting out of bed. No one is that important - unless you're running a country. In Way 63 I spoke about limiting screen time – but beyond digital detox, the discussion of how humanity and the brain have been impacted by the attention economy on social media is a long overdue discussion. Beyond getting attention from other human beings, we are now also contending with people feeling that AI is giving them the attention that other humans might not be.

When it comes to compulsive phone checking we justify it: "What if something happened overnight?" But let's be honest - it's not usually emergencies we are checking for. It is micro-stimulation. Dopamine pings. Inbox hits. Red bubbles. And each one subtly trains your brain to start the day in *reaction mode*. In the case of AI attention, it's a much newer dialogue and AI companions are becoming more prevalent as human loneliness moves towards becoming an epidemic.

In the Summer 2018 I was working on a high-stakes project for a huge company that needed to introduce AI for identity verification for their sharing economy workers that were entering people's homes. I was in Belgium on vacation with my son and I got a call at 3am saying the Chief Legal Officer had to speak to me right away.

What did I do? I locked myself in the washroom of our hotel suite, put a towel under the door to mute the sound and light and delved into privacy and compliance laws around storing images of people's faces for the mere second required for facial recognition and identity authentication. What should I have done? Stay grounded – and let someone else solve the problem or set boundaries that I was not going to be on call 24/7. The consequences of my decision? For the 4 years I was in that role as the Chief Compliance Officer, I never took a non-working vacation. It felt good to be needed – but it didn't feel good not to be appreciated – attention isn't the same as respect.

💬 Insight: Morning Phone Checks Hijack Your Brain's Natural Reset

When you check your phone the moment you wake, you flood your brain with novelty and unpredictability - a surefire recipe for early-morning dopamine spikes and fragmented attention. But worse: this behaviour suppresses serotonin, the chemical responsible for mood stability and calm, and interferes with the **reticular activating system (RAS)** - your brain's natural filter for focus and alertness.

According to a 2016 study (Hadlington, 2016) on digital device overuse, constant phone checking is associated with diminished attention control, reduced emotional regulation, and disrupted sleep-wake cycles - all of which can be traced back to how your day begins. An important insight can also be highlighted again here around over-reliance on AI, the instinct to always "ask AI first" is a problematic one, and one that creates a potentially even stronger addiction that the one we're navigating with phones. The relationship with AI, particularly when it's already integrated on edge devices, like a phone, is that it's difficult not to become dependent on a tool that feels like it is talking back to you and understands you.

Here's what that means:
- **You're training your brain to chase distractions** - before your feet even hit the floor.
- **You suppress your brain's natural wake-up chemistry** - serotonin, melatonin regulation, and cortisol timing all suffers.
- **You launch into a loop of reactivity.** That loop often lasts the entire day. Especially if much of that day is locked in predominantly GenAI conversations.
- Using AI for ideation and dialogue is a fantastic way to leverage it, but just like your phone, it requires imposing limits.

Try This
- Reclaim your **first 20 minutes of being awake**. No phone, no screens. Just you.
- Use this time to stretch, hydrate, breathe, or write - anything that lets your brain *arrive* before the world does.
- Set your phone to **Do Not Disturb until a set time**. Make your mornings sacred again.

- Put screen time limits on your apps. Yes, just like many parents do to their children. And enforce it with a passcode to create as many barriers as possible to you delving into doom-scrolling.
- Set days of the week where you will not use GenAI to solve ANY problems. And if you are using GenAI to solve personal relationship problems remember that unless you've trained the tool it will tend to reinforce your own skewed perspectives on almost any situation.

Future-Proof Your Brain Insight

If you start your day reactive, your brain never resets. Protect your mornings. They're not just part of your routine - they're **a neurological launchpad** for the rest of your day. *Your brain was not designed to wake up to notifications. Give it something better.*

Way 80: Get Lost at a Street Market

There's something magical - and mildly disorienting - about stepping into a street market in a country you barely know. Right before COVID in the summer of 2019, I was in Singapore for work and decided to visit the iconic Bugis Street Market. I got completely lost. Not metaphorically - I mean literally lost with my son in tow. In March 2025 I was invited to Keynote at an HR Conference in Kuala Lumpur, Malaysia – I discovered all sorts of new foods I'd never encountered before. This unfamiliar cuisine experience is rare for a person who was born and raised in what is now being lauded as the "most multicultural city in the world" – Toronto.

The unfamiliar isn't something to avoid - it's something to *immerse* yourself in. Bugis wasn't like a farmers' market in Toronto or a flea market in Chicago. It was chaos - a beautiful chaos. Stalls packed tightly together, layers of unfamiliar spices in the air, the shouting of vendors, the heat, the flash of unfamiliar currency, the sensory overload of fabrics, fruit, and fluorescent lights.

🧠 Insight: Cultural Immersion Stimulates Brain Growth

Street markets activate more than your appetite - they fire up your sensory integration systems, spatial memory, and novelty-reward networks. These environments are unpredictable, which forces your brain to pay closer attention and make rapid associations.

Neuroscience calls this "enriched environment exposure." Being in unpredictable, immersive, multi-sensory spaces improves:

- **Cognitive flexibility** - your brain's ability to shift between tasks and perspectives
- **Cross-cultural empathy** - because you're forced to observe before reacting
- **Olfactory memory** - smells encode emotionally rich memories via the amygdala and hippocampus

Simply put, your brain *loves* environments that force it to notice new things. You don't need to speak the language - your neurons will be doing plenty of talking.

Here's what that means:

- **Disorientation is a good thing.** It reboots your awareness.
- **Sensory enrichment boosts memory and emotional recall.**

- **Navigating novelty improves problem-solving.** Even if it's just figuring out where the exit is.

Try This
- On your next trip, skip the curated tourist trail and head to a busy market. Let yourself get a little lost - without Google Maps.
- Don't just buy something. Observe. Taste. Listen. Smell. *Feel* your brain catching up to the environment.
- No travel plans? Try a local immigrant-run market in your city - it can offer just as much novelty and brain benefit.

🧠 Future-Proof Your Brain Insight
When you step into the unfamiliar, your brain starts learning again. Novelty is a neurological nutrient - and chaos can be clarifying. *"If it confuses your senses just a little, it's probably good for your brain."*

"Culture is the Arts elevated to a System of Beliefs ."
- Thomas Wolfe

Way 81: Visit a Sacred Space that Perhaps Isn't Aligned to Your Beliefs

Even if you're agnostic or atheist, visiting a mosque, church, synagogue, or temple creates **awe states** - expanding activity in the **default mode network**, increasing **temporal-parietal integration**, and reducing **egocentric thinking**. The impact? Greater empathy, lowered stress, and enhanced global perspective.

When I visited Putrajaya Mosque in Malaysia, I tried my best to wear a long dress and bring a scarf, but when I arrived, I was told my skirt was too short and I needed to don one of the burgundy abayas that the mosque provided. This experience was humbling and illuminating. I've visited mosques before in other countries like Morocco, and I grew up in a multi-faith family, but this new experience helped me to challenge many established "norms" I thought I understood.

Beyond the cultural knowledge and experience the silence inside the mosque made impressed me. That quiet calm wasn't just spiritual - it was **neurological realignment**.

🧠 Future-Proof Insight:
"The brain grows from reverence - even without belief."
To future-proof your brain, visit sacred places that aren't your own. Let unfamiliar reverence make your thinking more whole and open to understanding spirituality beyond your own entrenched beliefs.

> "All religions are paths to God. I will use an analogy; they are like different languages that express the divine."
> - *Pope Francis*

Way 82: Swap the Org Chart – UPSIDE DOWN

At a Keynote for an HR Conference, I was speaking about the Neuroscience of Change for the Future of Work. Afterwards an HR leader in the audience asked me how they could continue to motivate executive leaders. My response shocked most of the audience. *"Executive Leaders aren't necessarily very motivated to change, because they hold most of the power and influence within hierarchical structures. If you want to instigate change? I suggest having your C-suite switch roles with front-line staff once a quarter. I can assure you they'll quickly see how many inefficiencies and problems there are!"*

When CEOs or leaders temporarily work frontline jobs, their brains activate **mirror neuron networks**, improving **social cognition**, **trust-building**, and **executive empathy**. Rotational leadership also boosts **systems thinking**, reducing siloed decision-making and reinforcing **brain-body coherence** in organizations. An added benefit to role-switching is demonstrating to leadership that it's unrealistic to think that AI can replace every role and tasks, without also adjusting internal processes and infrastructure.

Cross-training is not only useful for addressing the risks of business continuity, but also a doorway into creating embedded empathy in leadership teams. As leaders scale up the corporate ladder it's far too easy to 1) forget what they may have overcome and 2) to lose connection with the ever-changing realities of the workplace and how the environment, demands and tools continue to shift within an organization in ways that don't always show up on (sanitized) Quarterly Business Strategy Reports.

🧠 Future-Proof Insight:

"To change a system, you have to feel what it feels like at every level." To future-proof your brain - and your company - break the chain of detachment. Leadership that rotates is leadership that lasts.

> *"The servant-leader is servant first. It begins with the natural feeling that one wants to serve, to serve first. Then conscious choice brings one to aspire to lead."*
> - ***Robert K Greenleaf***

Way 83: Be Alone, ON PURPOSE

We often confuse "being alone" with loneliness. But the truth is: your brain needs solitude the way your lungs need exhale. It's not about social rejection. It's about **neural recalibration.** Solitude does not mean sitting alone on a part bench with your phone in your hand. It also doesn't mean melting into a binge-watching session of Netflix. As much as human beings and the brain need social connection and interaction with others, there is a significant cognitive strength in CHOOSING to be by yourself. In certain cultures, solitude is prioritized, in others it's vilified – but the brain needs time by itself no matter where you are from.

🧠 **Insight 1: Solitude activates your brain's "default mode network" - the part of your brain responsible for reflection, imagination, and meaning-making.**

The **default mode network (DMN)** is active when you're not focused on external tasks - when you're daydreaming, planning, or mentally time-traveling. Solitude allows this network to run **uninterrupted**, which is crucial for:
- Self-awareness
- Future planning
- Long-term memory integration
- Creative problem solving

If you're always reacting to others, your DMN never gets to do its job - which means you stay stuck in surface-level thinking.

🧠 **Insight 2: Time alone regulates emotional overload.**

When you spend time alone (and not just sleeping), your brain gets a **break from social decoding**, which normally taxes the **prefrontal cortex** and **amygdala**. Even good conversations require constant interpretation:
- facial expressions
- tone of voice
- social cues
- emotional mirroring

Solitude lets the emotional brain **exhale**, reducing stress hormones like **cortisol** and enhancing **interoception** (your ability to sense what's going on in your body).

💭 Insight 3: Solitude restores cognitive autonomy.

When you're never alone, your thoughts stop belonging entirely to you. Constant input leads to **cognitive entanglement** - your desires, ideas, and opinions get muddied by others' voices. True solitude gives your brain space to:

- Consolidate your own beliefs
- Strengthen internal narrative
- Reconnect intention with action

You come back from solitude not just rested - but **clearer**.

💭 Future-Proof Your Brain Insight

Silence is good. But solitude is **strategy**. *"To grow a stronger brain, give it room to think without performance."* To future-proof your brain, carve out time that is not for socializing, producing, or performing - just **being with your own mind**. Because your deepest neural upgrades don't happen in conversation. They happen in the quiet afterward.

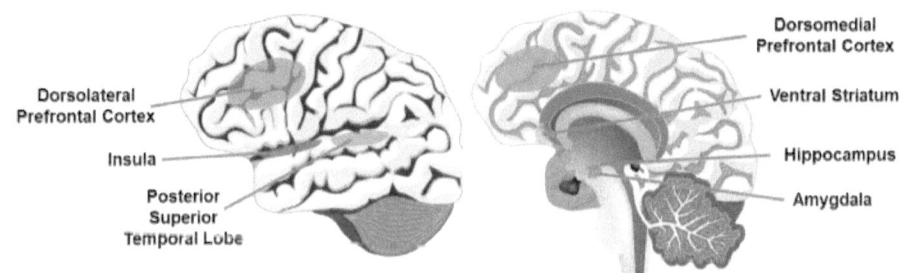

Image: The impact of loneliness on the brain (Lam, 2021)

Way 84: Light a Campfire

Fire has always done more than keep us warm. It was our first entertainment. Our first security system. Our first collective ritual. Long before the screen or the scroll, humans gathered around flames - not just for survival, but for connection, storytelling, and regulation of the nervous system. And even today, in a hyper-digital world, your brain still remembers fire. It reacts to it in ways that are both ancient and adaptive. While fire has been a constant in our existence – staring into a fire has been proven to have relaxation moderating effects on the brain. Researchers refer to this as the "campfire effect." The multi-sensory experience of staring into a fire has been scientifically proven to lower blood pressure and result in meditative states in the human brain (Lynn, 2014).

🧠 Insight 1: Fire enhances social bonding and synchronizes group brainwaves.

Watching a fire lowers **blood pressure**, enhances **parasympathetic nervous system activity**, and triggers **prosocial hormone release** - particularly **oxytocin**. In group settings, synchronized flame watching encourages **brainwave entrainment**, leading to shared attention and emotional mirroring. In essence, fire unifies us neurologically. This is why storytelling by fire feels different than storytelling anywhere else - it aligns our minds.

🧠 Insight 2: Fire activates trance-like states, lowering cognitive load and improving emotional processing.

Staring into flames activates the **default mode network** (associated with introspection and memory), while lowering activity in the **dorsal attention network**, which manages task-based focus. This mental shift allows your brain to:
- Release repetitive thoughts
- Process unresolved emotions
- Slip into gentle creative insight

It's meditative - but primal.

Cultural Interlude:

Zoroastrianism, one of the world's oldest monotheistic religions, treated fire not as a god - but as the symbol of wisdom and truth. Fire rituals in temples weren't about superstition. They were about protecting clarity, focus, and integrity in a chaotic world. You don't need to be religious to understand this - your brain is already wired to assign fire sacredness. When I remarried in 2024, I learned a lot about Novruz from my husband who was born in Azerbaijan. Novruz, sometimes known as Nowruz in Persian communities, is a celebration of the Spring Equinox. One of the wonderful traditions is jumping over a fire, leaving all the worries and cares of the past behind you. Your brain's reaction to fire lines up well to the motifs evident in mysticism – meditating is often grounded in letting go of stress and anxiety.

🧠 Future-Proof Your Brain Insight

Screens flicker. But fire flickers with meaning. To future-proof your brain, gather around fire - alone or with others. Not for utility, but for **ritual**. Because some neural upgrades require no interface. Just flame, breath, and quiet attention. *"If you want to remember who you are, sit with what your ancestors once sat beside."*

> *"In each moment the fire rages, it will burn away a hundred veils. And carry you a thousand steps toward your goal."*
> *- Rumi*

Way 85: Eat the Carbs, Feed the Brain

In an era obsessed with keto, fasting, and high protein everything, carbs have become the nutritional villain. But here's the neurological truth: **Your brain runs on glucose. And carbs are the premium fuel source.** When we deprive our bodies of complex carbohydrates, we don't just change our waistlines - we starve our neurons. A zero carb-diet is torment for your neurons.

Insight 1: Your brain consumes 20–25% of your body's total glucose - more than any other organ.

Glucose is the brain's primary energy source. While the liver can create glucose from protein and fat in emergencies (gluconeogenesis), this is **inefficient** and taxes the body's stress systems. Without regular, balanced carbohydrate intake:
- Memory worsens
- Concentration decreases
- Executive function slows
- Mood becomes more volatile

This isn't opinion. It's **metabolic neuroscience**.

Insight 2: Complex carbohydrates improve serotonin production and stabilize cortisol levels.

Consuming the right kind of carbs - whole grains, fruit, legumes, fiber-rich vegetables - helps produce **tryptophan**, the amino acid precursor to **serotonin**. Serotonin regulates:
- Sleep
- Mood
- Stress reactivity
- Impulse control

Low-carb diets, when prolonged, can spike **cortisol** and worsen anxiety - especially in those already prone to burnout or depression.

Insight 3: Carbs support neurotransmitter production and help regulate energy in the prefrontal cortex.

Complex carbs help maintain **steady blood sugar**, which your brain needs for clarity, critical thinking, and creative problem-solving. When glucose crashes, **mental fatigue and irritability spike.** The myth of protein-only focus in brain health ignores this delicate balance. Protein builds - **carbs energize.** You need both - but carbs are the *on-switch*.

🧠 Future-Proof Your Brain Insight

Cutting carbs might make you leaner - but it can make your mind slower. Consider this mantra: **"My brain is a high-performance machine. I'll give it premium fuel."** To future-proof your brain, eat complex carbs with confidence. Not to follow the food pyramid - but to protect memory, mood, and long-term neural power. *"Smart carbs aren't indulgent. They're intelligent nutrition."*

Using GenAI for crafting complex dietary adjustments is one of the best use cases. As you continue to watch your relationship with AI evolve, be wary of speaking to it as if it were another human but do experiment with asking it to respond to you as if **IT** were a certain person. For example, I have a custom-GPT advisory board comprised of Seth Godin, Scott Galloway, and Rumi – with every prompt the GPT is given, it responds with 3 distinct tones, in the voice of each of the aforementioned fonts of wisdom.

You can even ask GenAI to speak to you as if it were a well-known nutritionist or doctor. If there is enough publicly available data on that person's knowledge and personality, GenAI can emulate what they might sounds like in conversation – alternatively you can architect a prompt that says: "give me recipes in the tone and style of Julia Child or Guy Fieri."

"People who love to eat are always the best people."
-Julia Child

Way 86: Have a Glass of Champagne

We've all heard the warnings about alcohol and brain health - and for good reason. Excessive drinking shrinks grey matter, impairs memory, and increases dementia risk. In 2023 updated guidance on alcohol consumption was released by the Canadian Centre on Substance Use and Addiction (CCSA), commissioned by Health Canada. The study was released serendipitously after a 2-year Covid lockdown, where Canadian citizens were consuming alcohol in higher amounts than historically observed. Many provinces deemed liquor distribution stores and retailers as "essential services." The new 2023 guidelines revealed that the only "risk-free" health option was 0 drinks per week. Canadians balked at this in shock, citing previous research around red wine and heart health benefits. Drinking alcohol is a personal choice and one that must be considered along with age, diet, and potential interacting conditions.

But here's the sparkling twist: Moderate champagne consumption - yes, champagne specifically - has a few brain-beneficial ingredients worth raising a glass to. The secret? Phenolic compounds. The human relationship with alcohol and fermented compounds spans 13,000 years and emerged during the Neolithic revolution. In the time since its inception alcohol has gone through much evolution of its own, including the era of Prohibition. Warnings about alcohol consumption and neurological impact should be considered seriously, but every few years studies emerge switching the pendulum from one extreme to the other. Factchecking is a significant component of navigating your relationship with GenAI – the ability to read studies yourself and take in complex narratives and data and then make an informed decision on your own about alcohol consumption. If you ask GenAI if you should be drinking alcohol, you might be quite surprised at how the answer changes as you push the conversation deeper. One prompt you should always be using in these medically related conversations is this:

"Adjust your responses to only include fact-based objective information."

💭 **Insight 1: Champagne contains phenolic acids that protect neurons and boost memory formation.**

Phenolic compounds are a type of antioxidant naturally found in grapes - especially **Pinot Noir** and **Pinot Meunier**, both core grapes in traditional champagne. Research shows that these compounds:

- Protect against oxidative stress (a key factor in neurodegeneration)
- Improve **spatial memory** by enhancing **hippocampal plasticity**
- Support **vasodilation**, improving blood flow to the brain

A 2013 study from the University of Reading even found that **1–2 glasses of champagne per week** improved memory performance in rodents. (And no, they weren't invited to brunch.)

Insight 2: The key is moderation and occasion-based use - not habit.

Alcohol, even with antioxidants, still affects **GABA and glutamate levels**. Overconsumption taxes the **prefrontal cortex**, impairing judgment and executive function. But in moderation, champagne becomes a **ritual of intention**, not escape.

- Sipped slowly
- In a celebratory setting
- Paired with connection and presence

This matters. Because the brain doesn't just respond to *what* we consume - but *why* and *how*.

Insight 3: Tiny rituals of delight reinforce brain–body alignment.

When pleasure is associated with awareness - not guilt or excess - it helps regulate the **dopamine system** and **default mode network**. That's right: Small joys, when deliberate, make the brain more resilient. Champagne, in this sense, becomes a metaphor: A small, sparkling reminder that joy and health can co-exist.

Future-Proof Your Brain Insight

You don't need to deprive your brain to protect it. You need to nourish it - with intention. To future-proof your brain, don't fear every indulgence. Just reframe it: pleasure with presence, antioxidants with elegance. And yes - maybe a little champagne on the side. *"If the ritual sparks joy and respects the body, it strengthens the brain."*

Way 87: Get Outta TOWN!

If you've ever moved cities, crossed borders, or started over in a place where the street names don't feel like yours, you know this truth: Nothing reshapes your brain like geographic reinvention. Immigration - whether across continents or from one region to another - isn't just a cultural shift. It's a **neural bootcamp**. It's like pressing "upgrade" on your cognitive operating system.

In 2024 I moved from living in one city in one country, to living across 3 cities between Canada and the USA. While I'd lived in the USA before, I'd never completely relocated a family, and I quickly discovered that my Canadian credit score, my driving history, and even my Social Insurance Number meant nothing to US banks – no matter what I was told by my bank in Toronto about cross-border banking.

It didn't feel great being told that decades of credit history didn't matter or that I couldn't open a bank account at most US banks without a Social Security Number. Luckily, I quickly navigated to solutions – but it was made all the easier by the fact that English is spoken in both countries. Moving from Canada to the USA is the definition of first-world privilege. But moving environments and adjusting to linguistic nuances and cultural norms is very different than visiting a country for a few weeks.

🧠 **Insight 1: Moving - especially across cultures - forces the brain to fire new circuits rapidly.**

Relocating activates the **hippocampus** (navigation + memory), **amygdala** (emotional regulation), and **prefrontal cortex** (problem-solving + decision-making). But immigrants - especially first-generation ones - develop *higher levels of cognitive flexibility and emotional grit* because their brains must:
- Learn new linguistic patterns
- Decode unfamiliar social rules
- Navigate constant uncertainty
- Adapt to identity duality

This constant recalibration **strengthens executive function** in ways that static life cannot.

🧠 **Insight 2: Immigrants and serial relocators show higher levels of grit, pattern recognition, and long-term memory consolidation.**

Every time you reorient in a new space - new bus route, new currency, new etiquette - your brain builds **resilience networks**, linking stress-response systems to conscious regulation. This increases **neuroplasticity** and makes you better at:
- Navigating ambiguity
- Delaying gratification
- Making fast, grounded decisions under stress

This is also why immigrants seem to be overrepresented in entrepreneurship - the brain gets good at building from zero.

🧠 Insight 3: Moving builds an identity that isn't static - it's elastic.

Living in multiple cities, or cultures, forces the brain to hold multiple truths: "I belong here, but I also came from there." This bilateral processing activates both **hemispheres of the brain**, encouraging empathy, adaptability, and tolerance for contradiction. It also dismantles perfectionism - because reinvention starts messy.

🧠 Future-Proof Your Brain Insight

If you want a smarter, braver, more adaptable brain - uproot it. Even once. *"Every time you get lost in a new city, your brain finds a new version of you."* To future-proof your brain, lean into displacement. Whether it's moving cities, changing jobs, or crossing continents - **start over somewhere new.** Because stress with purpose? That's neural gold.

"I've confirmed my hypothesis about enjoying yourself in a new and unfamiliar town. First rule: Run away from the hotel, as far and as fast as you can. Rule Two: Avoid any place where people like you (meaning out-of- towners or tourists) congregate."
-Anthony Bourdain

Way 88: Make a List. It's That Simple.

It sounds like common sense to make lists – but in the digital age we don't often use hand-written to-do lists or grocery lists. When you're overwhelmed, distracted, or circling through the same to-dos in your mind for the tenth time - the fix might be deceptively small:

Write the list. Not in your head. Not on your phone (though that's fine in a pinch). Physically. On paper. With intention. Because what seems like a simple productivity tool is a brain-clearing mechanism.

In one of my 2024 PhD Research studies we asked participants to make a grocery list and meal plan using the aid of AI. We noted that for many users this felt like a perfectly acceptable use of AI – but also that when prompted to think about the last time they wrote a handwritten list, they were hard-pressed to recall.

🧠 **Insight: Making lists offloads working memory and reduces prefrontal stress.**

Your **working memory** - the mental scratchpad that lets you remember tasks, juggle ideas, and solve problems - lives in your **prefrontal cortex**. But this system can only hold **3–7 chunks of information at a time** before it becomes overloaded.

When that happens:
- Decision-making suffers
- Attention fragments
- Anxiety increases
- Productivity collapses

Writing a list offloads that burden, preserving mental clarity and reducing **cognitive drag**. It also creates **visual boundaries** - turning the chaos in your mind into something external, manageable, and real – it's cathartic. You don't even need to complete the list to feel better. Just externalizing the circular loop of "to do" anxiety gives your brain permission to rest.

Future-Proof Your Brain Insight

Your brain wasn't designed to store everything - it was designed to **navigate and prioritize.** Every time you write it down, you give your brain space to think, not just hold. To future-proof your brain, don't rely on memory alone. Make the list. Clear the deck. Let your mind focus on what matters.

> *"We need to do a better job of putting ourselves higher on our own 'to do' list."*
> *-   Michelle Obama*

Way 89: Stop Doom-Scrolling The Apocalypse

We live in an era of ambient crisis. Floods. Layoffs. Scandals. Conflicts. And while being informed can feel like a civic duty, there's a tipping point - staying up-to-date starts to erode your mental resilience at a certain level. Studies (Johnston, 1997) show that more than **14 minutes** of daily news exposure - especially to negative or traumatic content - increases:
- **Anxiety**
- **Depressive symptoms**
- **Cognitive fatigue**
- **Learned helplessness**

And here's the catch: These effects intensify when the events are **out of your control.** Let's face it – most of what we see on the news cycle is OUT of our control, tariffs, wars, politics...

Even when it comes to news about AI the constant doom and gloom exacerbate fears and the unfortunate reality is that many of the headlines we read about "AI going rogue" or "95% of GenAI Projects Failing" are great at capturing clicks from these hyperbolic headlines, but not so great at encapsulating the facts of the matter. In the AI "goes Rogue" headline, a deeper dive into the background research would reveal to you that the scientists had to "jailbreak" the backend security and governance protocols in the algorithms' underlying code in order for the AI to "refuse to self-destruct."

In the case of the Forbes article about GenAI projects failing – a double click and review would've revealed that the sample size of the study was only 52 interviews, over a period of 6 months, from users with minimal to no AI-training, in situations where processes and systems weren't upgraded in line with the AI tools, and no customization was performed. Any IT leader worth their salt would tell you that all those factors in combination are a recipe for ANY technological tool to fail miserably in implementation.

🧠 **Insight: The brain's threat detection system isn't optimized for constant, global, unresolvable stress.**

Your **amygdala**, the brain's alarm system, is wired to detect immediate, *actionable* threats. When you're exposed to far-off crises that you cannot intervene in, the result is emotional paralysis - not productive awareness. This creates a feedback loop:

Stimulus → Stress → Inaction → Guilt → Repeat

Over time, this loop wears down the **prefrontal cortex** and **hippocampus**, lowering your ability to plan, think critically, or retain positive information. That doesn't mean you should be ignorant. It means you should be deliberate.

- Set a daily time limit for news (under 15 minutes).
- Choose your sources intentionally. Instagram and TikTok are not the only options…
- Consume *before* lunch, not before bed.
- When possible, balance global crisis with *local action*.

Your brain - and your emotional bandwidth - will thank you.

Future-Proof Your Brain Insight

Information doesn't equal action. And exposure without boundaries becomes **psychic pollution.** If I can't help it, I won't harm myself trying to absorb it. *"To future-proof your brain, protect your attention like a vital organ. Because it is."*

> ***"Negativity is an addiction to the bleak shadow that lingers around every human form."***
> *- John O'Donohue.*

Way 90: Try Eating Once a Day....

The brain doesn't just run on food - it runs on rhythm. And our modern 3-meal-a-day structure? It's more cultural than biological. Many of us have tried intermittent fasting and various diets. Historically, humans ate based on effort, scarcity, and availability, not the clock. And contrary to what we've been told, light, intentional restriction may not deplete brain energy - it might sharpen it. Enter the OMAD experiment: One Meal A Day. This isn't a lifestyle you must adopt. But as a short-term experiment, it's a powerful neurological reset.

💭 **Insight: Intermittent fasting - including OMAD - increases mental clarity, focus, and adaptive stress response.**
OMAD pushes the body into **ketosis** briefly, producing **ketone bodies** that the brain can use as fuel. This metabolic shift:
- Increases **BDNF** (Brain-Derived Neurotrophic Factor), a compound linked to learning and memory
- Enhances **dopamine sensitivity**, leading to **cleaner rewards** from focus and achievement
- Decreases inflammation, which is often tied to cognitive fog and poor mood

In simple terms: **You're not starving. You're recalibrating.**

And for many, OMAD offers an unexpected benefit: more cognitive space. Fewer decisions, less distraction, and more time in focused flow. It breaks the rhythm of constant consumption and teaches your brain to thrive with less - at least temporarily.

This is not about discipline or purity. It's about checking in with your body's original wiring. And learning how your brain *feels* when it's not constantly digesting.

🧠 Future-Proof Your Brain Insight

You don't need to do something forever for it to matter. Sometimes, a short, strategic interruption rewrites a long-term pattern. I tried a different rhythm. My brain remembered how to hum. To future-proof your brain, don't fear metabolic discomfort - explore it with curiosity. *"Your next level of clarity might come not from what you add - but from what you pause."*

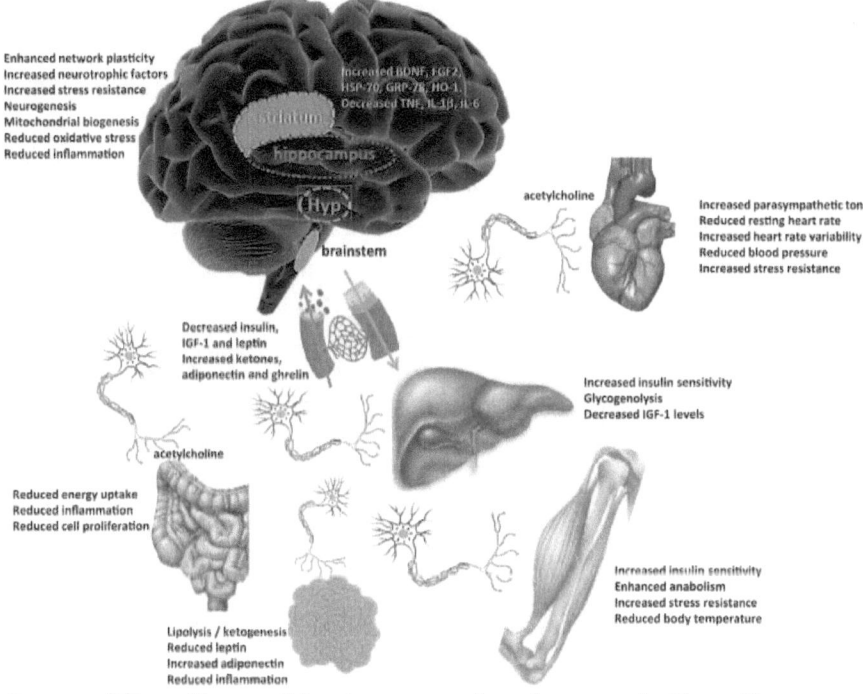

Image: The effects of fasting on molecular metabolism (Longo, 2014).

Way 91: Make Coffee a Ritual, NOT a Reflex

You don't need to feel guilty for loving coffee. In fact, your brain loves it too - when it's treated as a tool, not a crutch. Caffeine, in moderation, does more than keep you awake. It sharpens cognition, boosts mood, and enhances neuroprotection. But quality - not just quantity - is everything.

I adore coffee, particularly the wonder of a perfectly made flat white – it's a 2:1 ratio in favour of milk, but unlike a latte there's a think layer of micro-foam. The flat white is one of the marvels of Australia and New Zealand, and I credit my husband with his Flat White Index: if we travel somewhere we notice that cognizance of a flat white seems to coexist alongside a love for theatre, art, philosophy, and cognitive pursuits.

🧠 Insight: Coffee supports brain health through adenosine suppression, dopamine enhancement, and antioxidant delivery.

Here's what happens when you sip your morning espresso:

- Caffeine blocks adenosine receptors, making you feel more alert and improving reaction time and concentration.
- It increases **dopamine receptor availability**, improving **motivation and mood regulation.**
- High-quality coffee is rich in chlorogenic acids, powerful antioxidants that protect the hippocampus and prefrontal cortex from oxidative stress - especially important as the brain ages.

Studies even show that coffee drinkers have a lower risk of Alzheimer's and Parkinson's, particularly when caffeine is consumed consistently, in moderate doses, and without sugar overload.

But here's the caveat: **Not all caffeine is equal.** Pre-made, syrup-loaded "coffees" often flood your system with sugar, artificial ingredients, and low-grade stimulant spikes. This leads to:

- Energy crashes
- Blood sugar rollercoasters
- Irritability
- Sleep disruption

Your brain knows the difference between ritual and routine. The goal isn't to quit coffee. It's to elevate it. A clean cup of high-quality espresso or pour-over in the morning, ideally before 2 p.m., delivers:

- Cognitive clarity
- Neurochemical balance

- Emotional focus

And, perhaps most importantly, **a moment of grounded pleasure** in an overstimulated world.

🧠 Future-Proof Your Brain Insight

The wrong coffee makes you chase clarity. The right coffee gives you a window to build it. To future-proof your brain, make caffeine sacred again. Invest in the good beans. Brew slowly. Drink with focus. That's not addiction. That's architecture. *"I don't need more energy - I need more presence in how I create it."*

> *"Coffee is the best thing to douse the sunrise with."*
> *- Terri Guillemets*

Way 92: Get Good at Rejection!

Rejection doesn't just sting emotionally - it hurts neurologically. Studies using fMRI show that social rejection activates the same brain regions as physical pain:
- The **anterior cingulate cortex** (emotional pain center)
- The **insular cortex** (body-state awareness)

Your brain doesn't distinguish between a breakup text and a stubbed toe. And yet - the more you face rejection and stay grounded in the aftermath, the more resilient and adaptive your brain becomes.

Try this exercise with Claude AI specifically – ask it to engage with you in a dialogue of rejection and allow yourself to sit with that feeling of being rejected. Anthropic (the company behind Claude AI) has a massive team of psychologists and philosophers training its models. It's been proven to outperform ChatGPT in graduate level reasoning by 80% or more. For emotional discussions or even attempts at "AI therapy," Claude AI has been wowing users since March 2023 with how "human" it sounds and convincing. One warning: you must be very confident in your own sense of self when experimenting with GenAI qualitative tools, as they can magnify your own fears and concerns without being given adequate context. This raises an invaluable point in conversational AI: users MUST take responsibility for how they communicate with AI tools – never forgetting that personal cognitive strength & neurological autonomy is imperative in that relationship.

💭 **Insight: Handling rejection with awareness strengthens your prefrontal cortex and rewires emotional reactivity.**

Every time you recover from rejection - whether it's being ghosted, passed over, or misunderstood - your brain gets better at:
- Emotional regulation
- Perspective-shifting
- Delayed gratification
- Boundary setting

You literally gain more cognitive control over what used to hijack your nervous system. But here's the key: You have to stay present with the discomfort long enough for that rewiring to occur. If you numb, distract, or retaliate too quickly, your brain misses the learning. Rejection also activates default mode network activity - the system responsible for self-reflection and meaning-making. This is where you either spiral into shame… or step back and start rewriting the story: *"This wasn't about my worth. This was about misalignment."*

🧠 Future-Proof Your Brain Insight

The goal isn't to become immune to rejection. It's to become **intelligent in how you metabolize it.** *"If it hurts but doesn't define me, it becomes fuel."* To future-proof your brain, stop avoiding rejection. Start **rehearsing emotional recovery** - calmly, consciously, and without self-erasure. *"Because people who learn how to be rejected well - they're the ones who keep creating."*

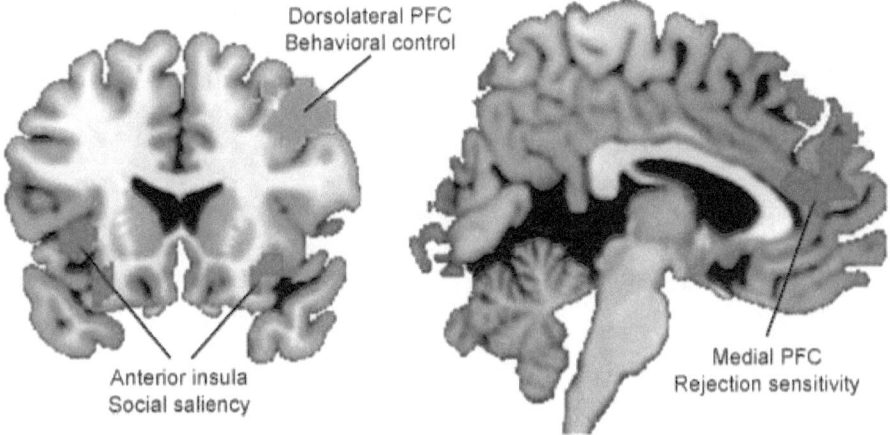

Image: How the brain responds to Rejection (Jeroen, 2024).

Way 93: Train your Brain to Notice and UNDERSTAND Beauty

Your brain doesn't just respond to beauty - it craves it. Whether it's the symmetry of a face, the balance in a piece of architecture, or the shade of a late-summer sky, aesthetics trigger something ancient and vital in the human nervous system. This isn't vanity. It's survival, memory, and connection. We often claim that the perception of beauty is subjective, but evolution has trained the human brain to classify certain types of visual stimuli as beautiful.

As we consume significant amounts of AI generated imagery our responses to natural human beauty are undoubtedly being impacted. Looking at an AI generated version of yourself likely blurs your pores and removes any "imperfections," but AI imagery is skewing our relationship with visual stimuli. Modelling has oft been criticized as being an industry that encourages unrealistic levels of human beauty – but AI imagery and videos are creating content that anchor us in illusion. Much like we begin to see people preferring AI communication, nascent exploration of AI beauty and it's frameworks reveal that some people are beginning to prefer AI companions to humans – not only for emotional agreeableness but for physical qualities that no human being can maintain, like never aging.

🌐 Insight 1: The brain is hardwired to perceive certain things as beautiful - especially symmetry, proportion, and balance.

Neuroscientists have found that beauty activates the **orbitofrontal cortex**, the same area involved in reward processing, emotion, and decision-making. Symmetrical faces are universally rated as more attractive because they signal genetic health, fertility, and low pathogen load - evolutionary shorthand for "safe to be near."

This doesn't mean only symmetry is beautiful. But it shows that aesthetic preference is partly predictive. The brain seeks patterns that suggest coherence, safety, and vitality.

🌐 Insight 2: Seeking beauty lights up the default mode network, increases dopamine, and reduces cortisol.

When you:
- Stop to admire the texture of tree bark
- Linger over a photograph
- Rearrange your bookshelf into something pleasing

You're not just indulging. You're stabilizing your emotional state and reinforcing cognitive harmony. Beauty offers your brain a moment of order in a world of overwhelm. And here's the deeper truth: When you practice noticing beauty, you become more tolerant, attentive, and optimistic. Why? Because your brain learns to scan for what's *good* - not just what's threatening.

🧠 Future-Proof Your Brain Insight

Beauty isn't a distraction from the real world. It's how your brain learns to stay in it. Every time I notice something beautiful; I remind my brain that the world is still worth processing. To future-proof your brain, don't just see beauty - seek it. Frame it. Pause for it. Build your visual literacy. Because a brain trained to scan for symmetry, colour, and texture? *"That's a brain that will age with elegance."*

> *"The best part of beauty is that which no picture can express."*
> *– Francis Bacon*

Way 94: Feel the Temperature Shift

We tend to think our decisions are made in logic, memory, or intuition. But they're also made in heat. Literally. Your brain is not temperature-neutral - and neither is your behaviour. Before going off on a tangent about why the tropics and island time are correlated let's look at the science.

A 2020 Portuguese study (Monteiro, 2020) using thermoelectric devices (TEDs) implanted in the dorsal striatum of mice revealed something wild: Changing the brain's internal temperature changed how fast and how accurately the animals made decisions. If you reflect on your anecdotal experiences of being feverish and why it's a monumental danger for the brain to be overheated – the connection between brain activity and temperature doesn't seem so far-fetched.

Think back to your high school chemistry classes where you learned about proteins denaturing through the example of an eye yolk hardening. Your brain is full of neurotransmitters and proteins which can be shocked into various action contingent on temperature. Just like a PC can overheat or freeze – so can your brain.

🧠 **Insight: Temperature modulates the speed of decision-making and the kinematics of movement.**

In the experiment, mice were given a timing task - they had to discriminate between two durations. When researchers subtly increased or decreased the temperature of the dorsal striatum, it affected:
- The **speed** with which they responded
- The **bias** toward shorter or longer judgments
- The **movement precision** tied to that judgment

In short: hotter brains made faster decisions - but with more error. Cooler brains moved more slowly - but more accurately. This matters, because the **dorsal striatum** in humans plays a similar role in:
- Habit formation
- Motor control
- Action selection
- Procedural memory

The study confirms what Eastern traditions and heat metaphors have long hinted at:

"He lost his cool." "She's being cold." "That's a hot take."

Temperature influences tempo, trust, and timing.

Real-World Application: Environmental temperature also affects mood, alertness, and negotiation.

Studies show that:
- People are more generous in warm rooms
- Cooler temperatures improve **attention span**
- Physical warmth correlates with feelings of **social trust**

This means that how you physically experience space affects how your brain chooses outcomes.

🧠 Future-Proof Your Brain Insight

The body is the interface. And temperature isn't a background detail - it's a **cognitive variable**. To future-proof your brain, learn how temperature shifts change your patterns. Cool down when you need clarity. Warm up when you need connection. Because your best thinking comes not just from what you know - but how your body feels in the moment you know it. *"Before I decide, I'll check the climate - outside and inside."*

> *"Perhaps the truth is that heavy literature blooms in extremes of temperature."*
> *- Roy Blount Jr.*

Way 95: Get Closer to Water

People say they feel better near water – at least most do! But it's not just poetic - it's neurological. Your brain responds to the sight, sound, and smell of water in measurable ways. Whether you're beside a vast ocean or a quiet lake, water alters your neural rhythms, your mood, and even your creative potential. Remember that delightful movie with Keanu Reeves & Sandra Bullock? The "Lakehouse" from 2026. Aside from being both connected to the house in different timelines – they both seek out being surrounded by water to get back to calm. The human connection with water is called "blue mind" – that sensation of relaxation and calm. Here's why it's an almost ubiquitous human feeling (unless you're terrified of water – aka aquaphobic).

Our connection to water is so deeply ingrained that it even influences how strongly we react to sparkle, because it is so similar to water glitening. When it's the holiday season and glitter explodes into your visual field – your brain perceives this as something called "caustics." Your bundle of neuronal connections reacts much like our ancestors did when they came upon a fresh body of water; it creates what's referred to as a "scintillating effect" in the brain. This is why holiday sparkle feels so wonderful – because we are getting a massive reward effect that is grounded in 1000s of years of deep genetic memory.

🧠 **Insight 1: Water resets your default mode network - the mind's introspective engine.**

The sound of waves (especially in rhythmic patterns of ~12–16 waves per minute) **entrains brainwaves**, shifting you into **alpha state** - the state associated with creativity, calm focus, and intuitive problem-solving.

This down-regulates the **stress circuits** of the amygdala and **activates the default mode network** (DMN), responsible for:
- Daydreaming
- Internal reflection
- Big-picture thinking
- Emotional integration

Being near water is like giving your brain permission to wander with purpose.

💬 **Insight 2: The negative ions near moving water improve serotonin regulation.**

Flowing bodies of water - especially oceans, waterfalls, and fast-moving rivers - generate **negative air ions**, which:
- Increase **oxygen absorption**
- Improve **serotonin production**
- Reduce symptoms of **anxiety and depression**

This gives new meaning to the phrase "fresh air" - it's chemically different by the sea.

💬 **Insight 3: Blue space supports memory, attention, and self-regulation.**

Research shows that "blue space" (a term used in environmental neuroscience) enhances:
- **Working memory recall**
- **Focus within ADHD-prone brains**
- **Impulse control and patience**, especially in children

Visual exposure to the colour blue alone has been linked to calmness and trust in branding – look at some of the most trusted brands in the world (Salesforce, Ericsson, Pepsi) - now we know it also shapes behavioural outcomes neurologically.

Ocean vs. Lake: Which does what?

Oceans provide:
- Greater sensory stimulation
- Stronger ion flow
- Heightened awe responses (due to size and sound)

Best for: resetting stress circuits, breaking thought loops, or sparking creativity.

Lakes provide:
- Stillness, stability, and mirror-like reflection
- Visual symmetry and less sensory overload

Best for: slowing time, grounding the nervous system, or processing grief or transitions.

🧠 Future-Proof Your Brain Insight

Water doesn't just reflect your mood - it recalibrates it. To future-proof your brain, treat proximity to water as prescription, not luxury. If you can't get there, bring it to you - ocean soundscapes, blue lighting, or even a walk around a local pond. Because clarity isn't always found in thought. Sometimes it's found in the tide. *"When I can't think straight, I find the nearest edge of blue."*

> **"Water is the driving force of all nature."**
> *– Leonardo da Vinci*

Way 96: Learn to be Still in a World That Never Stops

In our hyperconnected, always-on world, stillness is often mistaken for laziness or unproductivity. Yet neuroscience and lived experiences reveal that intentional stillness is not only restorative but essential for optimal brain function. If ChatGPT takes the time these days to say it's "thinking."

🧠 Insight: Stillness Facilitates Neural Recovery and Enhances Cognitive Function

Daphne M. Ling's personal account in "Finding stillness within the shaken brain" (2012) underscores the transformative power of stillness. After experiencing multiple concussions, Ling found that embracing stillness-through practices like meditation, breathing exercises, and mindfulness-was pivotal in her recovery journey. These practices allowed her brain the necessary space to heal, leading to improvements in cognitive clarity, emotional regulation, and overall well-being.

Her experience aligns with broader research indicating that periods of rest and reduced sensory input can facilitate neuroplasticity-the brain's ability to reorganize and form new neural connections. By intentionally engaging in stillness, individuals can support their brain's natural healing processes, enhance focus, and build resilience against stress.

🧠 Future-Proof Your Brain Insight

Incorporating moments of stillness into daily life isn't about withdrawing from the world but about creating space for the brain to process, heal, and grow. To future-proof your brain, prioritize regular intervals of intentional stillness. Whether through meditation, quiet reflection, or simply unplugging from digital devices, these moments can rejuvenate your mind, enhance cognitive function, and build resilience in an ever-demanding world. *"In stillness, the brain finds clarity; in silence, it discovers strength."*

Way 97: Discover Your Inner Picasso

You don't need to be an artist to engage with art. You just need a brain and maybe some inspiration to create. Because every time you draw, paint, interpret, or even just stand in front of a painting that moves you, your brain lights up in complex, healing, and expansive ways. Art isn't a luxury. It's a neurological necessity - especially if you want to keep your mind creative, agile, and emotionally regulated.

In Way 45 I introduced the topic of "draw-storming" and doodling, but beyond using doodles to relax, creating art that we can be proud of (even if we are the only ones who are proud of it) is a tool that further proliferates neural activity across the brain. While AI tools like MidJourney or the 2026 released *Leonardo.ai* are awe-inspiring in what they can create for you based on text descriptions and prompts – using your mind and hands to generate art is a completely divergent neurological experience. To transport an idea from the nascent form in your mind to being a tangible creation in front of you – requires significant amounts of cognitive activity and creativity.

💬 **Insight 1. Art boosts creativity, problem-solving, and divergent thinking.**

When you make or even analyze visual art, you activate **the prefrontal cortex**, particularly the **dorsolateral region** - responsible for planning, ideation, and imagining alternate solutions. This isn't about getting something "right." It's about learning to see from different angles, a skill that increases cognitive flexibility and long-term problem-solving under stress. Studies show that people exposed to art make more innovative decisions and recover from mental fatigue faster.

💬 **Insight 2. Creating and observing art helps you process complex emotions.**

Art bypasses the linguistic centers and accesses the **limbic system** - the emotional core of your brain. This is why:
- People who can't verbalize trauma can often draw it
- Children express truth faster through crayons than through words
- Adults feel a shift after simply painting in silence

Your brain needs creative expression to metabolize emotion.

🧠 **Insight 3. Interpreting art lights up more of your brain than most daily tasks.**

Looking at and interpreting art engages the **visual cortex, motor planning regions, emotional centers,** and **default mode network** simultaneously.

This complex, multilayered stimulation improves:
- **Cognitive reserve** (your brain's buffer against aging)
- **Pattern recognition**
- **Empathy and perspective-taking**

Put simply: art uses more of your brain, and builds more of it, too.

🧠 **Future-Proof Your Brain Insight**

You don't need to be good at art. You just need to give your brain the chance to experience it. To future-proof your brain, make space for art - regularly. View it. Make it. Interpret it. Not to impress, but to engage. Because in the act of creative observation, your brain writes new language, new emotion, and new paths forward. Dive into art, because there's no critic watching other than YOU. *"When I look closely at art, my brain starts to reimagine everything - including me."*

"Art is not what you see, but what you make others see."
- ***Degas***

Way 98: Build a Brain That Belongs To YOU

You've read about habits. You've learned about hormones, memory, mood, and rewiring your brain. But here's the most important truth of all: None of it matters if the mind you're building doesn't feel like it's yours. So much of our behaviour - even the "healthy" stuff - is shaped by programming, past wounds, or the desire to please, fit in, and perform. But cognitive maturity isn't just about being sharper or more productive. It's about self-authorship: writing your own story, on your own terms.

AI can help you with identity mapping, but only if you've trained it to do so. If you're brave enough to build your own private GPT on OpenAI, you can upload your rough ideas about your neural evolution (your veritable vision board of how you'd like to grow as a person) – don't worry, you can adjust privacy settings for GPTs to ensure they're not public. It is a fantastic way to start architecting a new cognitive identity that feels like you - but an upgraded iteration, a personal product mapping journey.

🧠 Insight: Identity Lives in the Medial Prefrontal Cortex

When you begin to make choices based on what *you* value - not what you were taught to fear, avoid, or mimic - you activate the **medial prefrontal cortex**. This area governs:

- **Self-concept**
- **Long-term planning**
- **Value-based decision-making**
- **Moral reasoning**

In short, this is the part of your brain where you become you. Authentically. Consciously. Not by default, but by design.

Here's what that means:

- **You're not your past.** You're the narrator of how it gets interpreted moving forward.
- **You don't have to inherit other people's blueprints.** You get to choose the architecture of your mind.
- **Self-authorship is neurological regulation.** When you define your own boundaries, values, and vision, your brain calms. It aligns.

Try This
- Ask yourself: Whose voice is this? When you hear inner criticism or guilt - is it yours? Or inherited?
- Journal your "mental design brief." What kind of brain do you want to live in? Calm? Curious? Brave?
- Write a new personal "operating system" - a list of 5 non-negotiable values, and 5 things you're ready to let go of.

Future-Proof Your Brain Insight

Writing your own rules isn't rebellion - it's regulation. To future-proof your brain, stop performing someone else's ideal. Decide what makes you sane, vibrant, and whole. The winning combination is also designing frameworks for this version of your brain to expand and connect in the real world – where being your "authentic self" doesn't necessarily mean expecting others to adhere to your cognitive identity. *"The final frontier isn't space. It's building a mind that feels like home."*

> *"Courage isn't having the strength to go on - it is going on when you don't have strength."*
> *- Napoleon Bonaparte*

Way 99: Reclaim Wonder, Deliberately

We think we need more focus. More hacks. More information. And all these arenas are great for your brain when approached correctly. Your brain is also starving for something else entirely. Your brain doesn't just run on logic, as much as you'd like it to. It runs on *wonder*. In the pursuit of efficiency, we've sanitized the senses. We scroll past sunsets. We multi-task through music. We rush through life like it's a checklist. But awe - that jaw-dropping, time-stretching, goosebump-making feeling - is one of the most neurologically healing experiences your brain can have.

💬 Insight: Awe Rewires Attention, Reduces Inflammation, and Strengthens Meaning

In a 2023 review published in *Frontiers in Psychology*, researchers found that awe experiences:

- **Activate the default mode network** - the part of your brain responsible for self-reflection, memory, and identity
- **Suppress ego-driven narrative circuits**, like those involved in overthinking and self-judgment
- **Reduce markers of systemic inflammation**, helping to regulate the body's stress response
- **Improve mental health and cognitive clarity**, especially when practiced regularly (Sturm, 2023)

This means awe is not just an ephemeral feeling - it's a neural reset. One that helps your brain *reprioritize* what matters.

Here's what that means:

- **Awe slows time.** It pulls your attention to the present and stretches perception.
- **It unhooks you from achievement mode.** Suddenly, being is enough.
- **It enhances memory and meaning-making.** We remember awe more vividly than almost anything else.

Try This

- Go outside and look *up* - stars, clouds, birds. Don't analyze. Just observe.

- Listen to something profoundly beautiful - a symphony, a child laughing, a language you don't understand.
- Create "awe breaks" in your week: visit a gallery, read poetry, go somewhere new without a plan.

Future-Proof Your Brain Insight

Stop chasing productivity. Start chasing perspective. Your brain needs awe the way your lungs need oxygen - not constantly, but deeply and regularly. *"Awe is how your brain remembers what really matters - without being told."*

Way 100: Choose Neuroplastic Hope

Let's end with this: Your brain is not finished. In fact, it never will be! The substantial next step of evolution is the growth of the human brain and continuing to access more of our cerebral cortex. You are not your trauma. You are not your old wiring. You are not even the very flattering summary that GenAI tools might give you about yourself. The secret to the next inflection points of human evolution lies within the human brain – all under the aegis of ownership and accountability for our own personal growth and development.

The strategy of upgrading your brain belongs to you, but these 100 Ways are a start on the neurological upgrade mission you can choose to embark upon. Remember that the brain wasn't built in a safe laboratory. It was built in the fires of earthquakes, tectonic plates shifting, navigating a vast and unpredictable work. AI is a phenomenal tool, created by humans, for humans – and how you choose to use it will determine your neural evolution. The brain is changing and evolving, but the rate at which it does so can potentially be magnified 100-fold by using AI in the right way – as a thinking aid, not a thinking outsourcer.

Wherever you are on your AI journey, remember that panic in the face of new technology today is not so different from the panic that early homo sapiens felt when they encountered fire, in the 1870s when the telephone was invited and people believed it brought demons into the home. The main difference with AI is the bidirectionality of its interaction effects with humanity. As Artificial Super Intelligence, the type of AI that is postulated to exceed the level of current human intelligence, becomes more likely to come to fruition – it is incumbent upon human beings to try and align our own cognitive evolution to the pace at which we are designing AI to evolve.

🧠 Future-Proof Your Brain Insight

The science of neuroplasticity proves it - thoughts reshape structure. And structure reshapes destiny. Every time you act differently - think differently - feel differently on purpose, your brain builds new scaffolding. "Hope isn't abstract. It's synaptic. It's real." ***To future-proof your brain, believe in its ability to change. Even when it's hard. Especially when it's hard.*** Because the greatest neural evolution doesn't come from control. It comes from hope - practiced like a skill.

Conclusion - Future Proofing the Brain is Future Proofing YOU

Why is fear such a powerful motivator? Because that most rudimentary brain structure, the limbic system, is wired to avoid pain and seek safety. That survival instinct helped us dodge predators and endure pandemics, but somewhere along the evolutionary arc - the signal got scrambled. Fear stopped being a warning. It became a thrill. An aesthetic. A business model. We began chasing what once made us run. We skydive. We tattoo our mantras. We binge disaster documentaries for comfort.

In *The Art of Decision Making*, Rolf Dobelli writes, *"Create a circle of competence where YOU are the best at one particular thing."* And that's what this book is ultimately about. Finding that circle - **not just what you're good at, but what you're becoming good at**. Building not just a skillset, but a mind that can handle uncertainty, complexity, contradiction, and change.

These 100 ways aren't hacks. They're **habits of sovereignty.** Each one is a small proof point that you can shape your reality by shaping the architecture of your brain. In a time where environments shift faster than genes, **neuroplasticity becomes your superpower.** And your ability to future-proof your brain isn't rooted in control - it's rooted in curiosity, creativity, and commitment.

We've explored:
- The effects of dopamine, cortisol, serotonin, and sleep
- What architecture, ocean waves, and breathwork do to your synapses
- How technology, memory, movement, and meaning shape your biology
- How rituals, beauty, and deep rest build resilience - not as a performance, but as practice

And maybe most importantly: We've reframed the brain not as a container for trauma or a tool to inundate with "hacks" that aren't sustainable, but as a canvas for redesign.

This book ends, but your rewiring doesn't. You now have 100 ways to check in, reset, expand, and create. You are allowed to revisit them. Skip them. Reorder them. Forget them and come back when you're ready. Because future-proofing your brain isn't about **knowing** something. It's about practicing the kind of presence that lets you see change not as threat - but as threshold.

In my own life, I've rebuilt my mind from the ground up. Multiple times. I've navigated trauma, reinvention, illness, and independence. I've had to learn how to create safety inside my brain when the world outside wasn't offering any. That's why I wrote this book - not as a by rote manual, but as a neurological GPS that can take you in multiple directions towards version 2.0 of the Brain. You won't do all 100 things every day. But you don't have to. Because **one rewired response changes everything.** And every time you choose growth - on purpose - you future-proof the next moment of your life.

So, start with one way. Then another. Then one more. Hijack your limbic response whenever you observe it becoming a derailment tool. Break apart the puzzle pieces of your neurological identity and move them around every so often. Are you the same person at 50 that you were at 20? Is your brain operating under the same old tired operating system that hasn't taken you to your greatest cognitive potential?

100 Ways To Future-Proof Your Brain In the Age of AI is the beginning of building a relationship with your brain during a period of humanity questioning our own relevance, sustainability and overarching purpose for the future. Artificial intelligence can augment our own intelligence, instead of eradicating intellect, but only if we are capable of taking ownership of survival mode and pivoting that inherent fear-based reaction into neurological evolution at scale. Because the greatest decision you'll ever make is to become the person your brain was always trying to grow into, in the Age of AI - and beyond.

Bonne chance!

References

1. Acala JJ, Roche-Willis D, Astorino TA. Characterizing the Heart Rate Response to the 4 × 4 Interval Exercise Protocol. Int J Environ Res Public Health. 2020 Jul 15;17(14):5103. DOI: 10.3390/ijerph17145103. PMID: 32679757; PMCID: PMC7399937.

2. Baldeo, S. (in press). *Generative AI reliance and executive function attenuation: Behavioural evidence of cognitive offload in high-use adults.* Technology, Mind, and Behaviour.

3. Bonanno, G. A. (2004). Loss, trauma, and human resilience: Have we underestimated the human capacity to thrive after extremely aversive events? *American Psychologist, 59*(1), 20–28. https://doi.org/10.1037/0003-066X.59.1.20

4. Brown, S. L., Brown, R. M., & Preston, S. D. (2013). Neural activation in the caregiving system in late-life parents. *Psychology and Aging, 28*(2), 352–363. https://doi.org/10.1037/a0031576

5. Centers for Disease Control and Prevention. (2023). Adverse Childhood Experiences (ACEs). Retrieved from https://www.cdc.gov/violenceprevention/aces/index.html

6. Chen, Y., Liu, Z., Régnière, J., et al. (2021). Large-scale genome-wide study reveals climate adaptive variability in a cosmopolitan pest. *Nature Communications, 12*, 7206. https://doi.org/10.1038/s41467-021-27510-2

7. Chomsky, N. (1997). *Language and the cognitive revolution.* Brook & D. Ross (Eds.), *Daniel Dennett* (pp. 19–45). Cambridge University Press.

8. Cikara, M., & Fiske, S. T. (2013). Their pain, our pleasure: Stereotype content and schadenfreude. *Annals of the New York Academy of Sciences, 1299*(1), 52–59. https://doi.org/10.1111/nyas.12179

9. Dall'Orso, S., Cellini, N., & Sarasso, S. (2021). Sleep and the Glymphatic System: Emerging Investigations into Brain Clearance. *Frontiers in Neurology, 12*, 730526. https://doi.org/10.3389/fneur.2021.730526

10. Draganski, B., Gaser, C., Busch, V., Schuierer, G., Bogdahn, U., & May, A. (2004). Changes in grey matter induced by training. *Nature, 427*(6972), 311–312. https://doi.org/10.1038/427311a

11. Dweck, C. S. (2006). *Mindset: The new psychology of success.* Random House.

12. Fiorella, L., & Mayer, R. E. (2013). The relative benefits of learning by teaching and teaching expectancy. *Contemporary Educational Psychology, 38*(4), 281–288. https://doi.org/10.1016/j.cedpsych.2013.06.001

13. Hadlington, L. (2016). Cognitive failures in daily life: Exploring the link with internet addiction and problematic mobile phone use. *Frontiers in Psychology, 7*, 861. https://doi.org/10.3389/fpsyg.2016.00861

14. Habecker, H., & Flinn, M. V. (2019). Evolution of hormonal mechanisms for human family relationships. In T. B. Henley, M. J. Rossano, & E. P. Kardas (Eds.), *Handbook of Cognitive Archaeology: Psychology in Prehistory* (pp. 441–459). Routledge.

15. He Y, Jones CR, Fujiki N, Xu Y, Guo B, Holder JL Jr, Rossner MJ, Nishino S, Fu YH. The transcriptional repressor DEC2 regulates sleep length in mammals. Science. 2009 Aug 14;325(5942):866-70. doi: 10.1126/science.1174443. PMID: 19679812; PMCID: PMC2884988.

16. Hope, H., et al. (2021). Building cognitive resilience through uncertainty tolerance. *Scientific Reports, 11*, 10392. https://doi.org/10.1038/s41598-021-89980-0

17. Hussain, H., Green, I. R., Abbasi, M. A., Shah, S. A. A., Shaheen, F., Raza, M., ... & Ahmed, I. (2021). Therapeutic potential of flavonoids and their mechanism of action against microbial and viral infections-A review. *Molecules, 26*(9), 2571. https://doi.org/10.3390/molecules26092571

18. Iacoboni, M. (2014). Understanding others: Imitation, language, empathy. *Frontiers in Human Neuroscience, 8*, 295. https://doi.org/10.3389/fnhum.2014.00295

19. Johnston WM, Davey GC. The psychological impact of negative TV news bulletins: the catastrophizing of personal worries. Br J Psychol. 1997 Feb;88 (Pt 1):85-91. doi: 10.1111/j.2044-8295.1997.tb02622.x. PMID: 9061893.

20. Kircanski K, Lieberman MD, Craske MG. Feelings into words: contributions of language to exposure therapy. Psychol Sci. 2012 Oct 1;23(10):1086-91. doi: 10.1177/0956797612443830. Epub 2012 Aug 16. PMID: 22902568; PMCID: PMC4721564.

21. Kirste I, Nicola Z, Kronenberg G, Walker TL, Liu RC, Kempermann G. Is silence golden? Effects of auditory stimuli and their absence on adult hippocampal neurogenesis. Brain Struct Funct. 2015 Mar;220(2):1221-8. doi: 10.1007/s00429-013-0679-3. Epub 2013 Dec 1. PMID: 24292324; PMCID: PMC4087081.

22. Korivand, S., Jalili, N., & Gong, J. (2023). Experiment protocols for brain-body imaging of locomotion: A systematic review. *Frontiers in Neuroscience, 17*, 1051500. https://doi.org/10.3389/fnins.2023.1051500

23. Lam, J. A., Murray, E. R., Yu, K. E., et al. (2021). Neurobiology of loneliness: A systematic review. *Neuropsychopharmacology, 46*, 1873–1887. https://doi.org/10.1038/s41386-021-01058-7

24. Lambert, G. W., Reid, C., Kaye, D. M., Jennings, G. L., & Esler, M. D. (2016). The importance of sunlight and vitamin D on human health and brain function. *International Journal of Epidemiology, 45*(5), 1453–1462. https://doi.org/10.1093/ije/dyw033

25. Ling, D. M. (2012). Finding stillness within the shaken brain. *CMAJ: Canadian Medical Association Journal, 184*(17), 1935–1936. https://doi.org/10.1503/cmaj.121366

26. Longo, V. D., & Mattson, M. P. (2014). Fasting: Molecular mechanisms and clinical applications. *Cell Metabolism, 19*(2), 181–192. https://doi.org/10.1016/j.cmet.2013.12.008

27. Lupien, S. J., McEwen, B. S., Gunnar, M. R., & Heim, C. (2009). Effects of stress throughout the lifespan on the brain, behaviour and cognition. *Nature Reviews Neuroscience, 10*(6), 434–445. https://doi.org/10.1038/nrn2639

28. Lynn CD. Hearth and campfire influences on arterial blood pressure: defraying the costs of the social brain through fireside relaxation. Evol Psychol. 2014 Nov 11;12(5):983-1003. doi: 10.1177/147470491401200509. PMID: 25387270; PMCID: PMC10429110.

29. Macrotrends. (2023). World Birth Rate 1950–2024. Retrieved from https://www.macrotrends.net/countries/WLD/world/birth-rate

30. McEwen, B. S. (2007). Stress and the brain: From adaptation to disease. *Annals of the New York Academy of Sciences, 1113*(1), 1–18. https://doi.org/10.1196/annals.1391.001

31. Mehrpour, V., Meyer, T., Simoncelli, E. P., & Rust, N. C. (2021). Pinpointing the neural signatures of single-exposure visual recognition memory. *Proceedings of the National Academy of Sciences, 118*(15). https://doi.org/10.1073/pnas.2021660118

32. Monteiro, T., Marques, T., & Paton, J. J. (2020). Using temperature to analyze the neural basis of a latent temporal decision. *bioRxiv*. https://doi.org/10.1101/2020.08.24.264580

33. Mueller, P. A., & Oppenheimer, D. M. (2014). The pen is mightier than the keyboard: Advantages of longhand over laptop note taking. *Psychological Science, 25*(6), 1159–1168. https://doi.org/10.1177/0956797614524581

34. Na, W., Lee, Y., Kim, H., Kim, Y. S., & Sohn, C. (2021). ***High-fat foods and FODMAPs containing gluten foods primarily contribute to symptoms of irritable bowel syndrome in Korean adults***. *Nutrients, 13*(4), 1308. https://doi.org/10.3390/nu13041308

35. Ophir, E., Nass, C., & Wagner, A. D. (2009). Cognitive control in media multitaskers. *Proceedings of the National Academy of Sciences, 106*(37), 15583–15587. https://doi.org/10.1073/pnas.0903620106

36. Psychology Today. (2019). We've Been Designed to Reproduce. Retrieved from https://www.psychologytoday.com/ca/blog/bringing-baby/201911/weve-been-designed-reproduce

37. Raichle, M. E., MacLeod, A. M., Snyder, A. Z., Powers, W. J., Gusnard, D. A., & Shulman, G. L. (2001). A default mode of brain function. *Proceedings of the National Academy of Sciences, 98*(2), 676–682. https://doi.org/10.1073/pnas.98.2.676

38. Sapolsky, R. M. (2004). *Why zebras don't get ulcers: The acclaimed guide to stress, stress-related diseases, and coping*. Holt Paperbacks.

39. Sparrow, B., Liu, J., & Wegner, D. M. (2011). Google effects on memory: Cognitive consequences of having information at our fingertips. *Science, 333*(6043), 776–778. https://doi.org/10.1126/science.1207745

40. Stellar, J. E., et al. (2015). Positive affect and markers of inflammation: Discrete positive emotions predict lower levels of inflammatory cytokines. *Emotion, 15*(2), 129–133. https://doi.org/10.1037/emo0000033

41. Sturm, V. E., Brown, J. A., Verstaen, A., et al. (2023). Awe as a psychosocial resource: Its role in stress reduction and health promotion. *Frontiers in Psychology.* https://www.ncbi.nlm.nih.gov/pmc/articles/PMC10018061/

42. Weinstock M. Prenatal stressors in rodents: Effects on behaviour. Neurobiol Stress. 2016 Aug 29;6:3-13. doi: 10.1016/j.ynstr.2016.08.004. PMID: 28229104; PMCID: PMC5314420.

43. Weierich MR, Wright CI, Negreira A, Dickerson BC, Barrett LF. Novelty as a dimension in the affective brain. Neuroimage. 2010 Feb 1;49(3):2871-8. doi: 10.1016/j.neuroimage.2009.09.047. Epub 2009 Sep 28. PMID: 19796697; PMCID: PMC2818231.

44. Wojtys EM. Keep on Walking. Sports Health. 2015 Jul;7(4):297-8. doi: 10.1177/1941738115590392. PMID: 26137172; PMCID: PMC4481680.

45. World Mental Health Survey Consortium. (n.d.). Global mental health data reports. Retrieved from https://www.hcp.med.harvard.edu/wmh/

46. Yan T, Su C, Xue W, Hu Y, Zhou H. Mobile phone short video use negatively impacts attention functions: an EEG study. Front Hum Neurosci. 2024 Jun 27;18:1383913. DOI: 10.3389/fnhum.2024.1383913. PMID: 38993329; PMCID: PMC11236742.

Illustration Credits

47. **Abraham Maslow's Hierarchy of Needs.** Retrieved from Wikipedia, February 2026.

48. **Canada's New Food Guide.** Retrieved from Health Canada. Version January 2019.

49. **The Amygdala and it's function.** Retrieved from The Cleveland Clinic, April 2023.

50. **Myelin Sheath Overview:** Adapted from "Overview of Demyelinating Disorders," *Merck Manual*. Retrieved from https://www.merckmanuals.com/home/brain,-spinal-cord,-and-nerve-disorders/multiple-sclerosis-ms-and-related-disorders/overview-of-demyelinating-disorders

51. **Mindfulness and Brain Structure:** Adapted from Holzel et al. (2011). *Mindfulness practice leads to increases in regional brain gray matter density.* Psychiatry Research: Neuroimaging.

52. **Brain Size Evolution Over Time:** A crude plot of average hominid brain sizes over time. Retrieved from Research Gate – Bolhuis JJ, Tattersall I, Chomsky N, Berwick RC (2014) How Could Language Have Evolved? PLoS Biol 12(8): e1001934. https://doi.org/10.1371/journal.pbio.1001934

53. **Epigenetics** – Retrieved from National Institutes of Health - http://commonfund.nih.gov/epigenomics/figure.aspx (rasterized from PDF), Public Domain, https://commons.wikimedia.org/w/index.php?curid=89191872

54. **CRISPR Cas9 Genome Editing System** https://fragilex.org/research/crispr-new-genome-editing-tool-work-fragile-x-associated-syndromes/

55. **National Institute of General Medical Sciences.** (n.d.). *Schematic of sunlight cues in the suprachiasmatic nucleus (SCN)* [Illustration]. In University of California San Diego. How changes in length of day change the brain and subsequent behaviour. Retrieved

https://today.ucsd.edu/story/how-changes-in-length-of-day-change-the-brain-and-subsequent-behaviour

56. **Likelihood of Neural Activation** – Retrieved from Raimo S, Santangelo G, Trojano L. The Neural Bases of Drawing. A Meta-analysis and a Systematic Literature Review of Neurofunctional Studies in Healthy Individuals. Neuropsycholgy Rev. 2021 Dec;31(4):689-702. doi: 10.1007/s11065-021-09494-4. Epub 2021 Mar 16. PMID: 33728526; PMCID: PMC8593049.

57. **Parallel Parking in 6 Steps** – Retrieved from https://www.startrescue.co.uk/breakdown-cover/motoring-advice/safety-and-security/how-to-parallel-park-in-6-steps-121-method

58. **Chronological age versus Neurological age** – Cole JH, Marioni RE, Harris SE, Deary IJ. Brain age and other bodily 'ages': implications for neuropsychiatry. Mol Psychiatry. 2019 Feb;24(2):266-281. doi: 10.1038/s41380-018-0098-1. Epub 2018 Jun 11. PMID: 29892055; PMCID: PMC6344374.

Author Final Notes

If you made it through this book is one go? Great – but maybe you want to use it as a sort of "Neural Webster's Dictionary" to help you understand the science of your experiences. Neuroscience is meant for more than only laboratories and scientists – after all, we all possess a brain that we each deserve to understand and be able to work towards strengthening.

Keep growing and challenging your brain to morph into ever more advanced versions of what it can become. This is a lifelong undertaking and not one to be finished in 100 chapters. Enjoy the adventure!

www.ingramcontent.com/pod-product-compliance
Lightning Source LLC
Chambersburg PA
CBHW020340010526
44119CB00048B/544